Bruno Dietrich

Gene expression profiling in differently fed farm animals

Bruno Dietrich

Gene expression profiling in differently fed farm animals

Effects of selenium and deoxynivalenol on the gene expression in the liver of broilers. Implementation of nutrigenomics.

Südwestdeutscher Verlag für Hochschulschriften

Impressum/Imprint (nur für Deutschland/only for Germany)
Bibliografische Information der Deutschen Nationalbibliothek: Die Deutsche Nationalbibliothek verzeichnet diese Publikation in der Deutschen Nationalbibliografie; detaillierte bibliografische Daten sind im Internet über http://dnb.d-nb.de abrufbar.
Alle in diesem Buch genannten Marken und Produktnamen unterliegen warenzeichen-, marken- oder patentrechtlichem Schutz bzw. sind Warenzeichen oder eingetragene Warenzeichen der jeweiligen Inhaber. Die Wiedergabe von Marken, Produktnamen, Gebrauchsnamen, Handelsnamen, Warenbezeichnungen u.s.w. in diesem Werk berechtigt auch ohne besondere Kennzeichnung nicht zu der Annahme, dass solche Namen im Sinne der Warenzeichen- und Markenschutzgesetzgebung als frei zu betrachten wären und daher von jedermann benutzt werden dürften.

Coverbild: www.ingimage.com

Verlag: Südwestdeutscher Verlag für Hochschulschriften GmbH & Co. KG
Heinrich-Böcking-Str. 6-8, 66121 Saarbrücken, Deutschland
Telefon +49 681 37 20 271-1, Telefax +49 681 37 20 271-0
Email: info@svh-verlag.de

Zugl.: Zürich, ETH, Diss., 2011

Herstellung in Deutschland (siehe letzte Seite)
ISBN: 978-3-8381-3206-8

Imprint (only for USA, GB)
Bibliographic information published by the Deutsche Nationalbibliothek: The Deutsche Nationalbibliothek lists this publication in the Deutsche Nationalbibliografie; detailed bibliographic data are available in the Internet at http://dnb.d-nb.de.
Any brand names and product names mentioned in this book are subject to trademark, brand or patent protection and are trademarks or registered trademarks of their respective holders. The use of brand names, product names, common names, trade names, product descriptions etc. even without a particular marking in this works is in no way to be construed to mean that such names may be regarded as unrestricted in respect of trademark and brand protection legislation and could thus be used by anyone.

Cover image: www.ingimage.com

Publisher: Südwestdeutscher Verlag für Hochschulschriften GmbH & Co. KG
Heinrich-Böcking-Str. 6-8, 66121 Saarbrücken, Germany
Phone +49 681 37 20 271-1, Fax +49 681 37 20 271-0
Email: info@svh-verlag.de

Printed in the U.S.A.
Printed in the U.K. by (see last page)
ISBN: 978-3-8381-3206-8

Acknowledgment

I am deeply grateful to all those who contributed to this PhD thesis.
For giving me the opportunity to accomplish this thesis I am very grateful to Prof. Dr. Caspar Wenk. I would also like to thank him for his continuous support and the inspiring discussions.
I would like to thank Prof. Dr. Peter Spring from the Swiss College of Agriculture in Zollikofen for the challenging project proposals and the donation of the corresponding feed materials. In addition I like to thank him for the important and rapid procurement between my work and the sponsor Alltech Inc. (Lexington, Kentucky) at all times of the day.
Many thanks to PD Dr. Stefan Neuenschwander for his scientific advices and the offering of very constructive suggestions in formulating this thesis.
I am very grateful to the team from Alltech Inc. (Lexington, Kentucky, USA) for giving me the possibility to carry out my PhD project. I like to specially mention the founder and President of Alltech Dr. Pearse Lyons for his passion for science and the support of young scientists. I am very grateful to the Director of Global Research Alltech Prof. Dr. Karl Dawson and the Director of Research Alltech Dr. Ronan Power for his advices and the possibility to present my results at different occasions.
I am particularly grateful to Prof. Dr. Wolfgang Langhans, Physiology and Behaviour Laboratory, Institue of Food, Nutrition and Health, ETHZ for agreeing to be my examiner.
A big thank to Dr. Hannes Jörg for introducing me into the field of molecular genetics and the scientific project discussions.
Sincere thanks go to Prof. Dr. Peter Vögeli for the help with the blood sampling and the helpful advices.
I thank Dr. Ruth Messikommer for her support in the animal trials, the administrative work and especially for encouragement during some difficult periods.
I thank Dr. Benjamin Bucher for the help with the launch of my PhD thesis and the slaughtering.
I like to thank Monika Haubitz and Dr. David Joller for all kind of help.
For the help with the fatty acid analysis I like to thank Eva Katharina Richter and Dr. Florian Leiber.
I acknowledge Carmen Kunz, Monika Büchel and Rolf Bickel for their help in the laboratory and the care of the animals.
For excellent co-operation special thanks to all my colleagues and friends at the institute, who contributed in many different ways to the accomplishment of this work: Dr. Aldona Pienkowska-Schelling, PD Dr. Claude Schelling, Benita Pineroli, Elisabeth Wenk, Gerda Bärtschi, Urs Schuler, the PhD students Antonio Rampoldi and Benjamin Furter.
Big thanks to Catharine Aquino and Andrea Patrignani from the Functional Genomics Center Zurich for the help with the hybridization and scanning of the microarrays.
I am most grateful to my family and my wife Patricia, who always supported my ideas and encouraged me during my time as a PhD student.

Contents

Summary

The study of the effects of a feed supplement or a contaminant on metabolites, protein activities and amounts in an organ is time consuming and requires comprehensive knowledge about cellular processes that may be influenced. Therefore, gene expression analysis was performed in the liver of broilers to screen for differently expressed genes at the level of mRNA. The liver was used because of its central role in metabolism and detoxification. In addition, the liver is exposed to all the digested nutrients absorbed into the portal vein. Chicken Genome Arrays (Affymetrix, Santa Clara, CA, USA) were used for the gene expression analysis, containing 32'000 probe sets, which correspond to 28'000 transcripts. The expression fold change (fc) between the groups had to be higher than 1.2 and the p-value had to be below 0.05. In three projects the gene expression in the liver of broilers was determined depending on the addition of selenium and different concentrations of Deoxynivalenol (DON) to the feed.

For the evaluation and establishment of the gene expression technique the animals from the first project were fed with a feed supplemented with selenized yeast (Se-yeast). Se-yeast consists mainly of L-selenomethionine, which can also be incorporated non-specifically into proteins during synthesis. Selenium is an essential micronutrient and a cofactor of the antioxidant enzyme glutathione peroxidase (GPX). In the animal trial 16 broilers per group were fed either a control diet or a diet supplemented with selenium (0.5mg/kg feed) for 35 days.

Totally 104 genes were significantly upregulated and 349 genes were significantly downregulated in the Se-yeast group. The selenium status of the birds in the Se-yeast group was increased, which was shown by the 2-fold increase of the GPX activity in the heparinized blood plasma. Beside the activity increase also the expression of *GPX1* (fc: 4.16) and *GPX4* (fc: 1.43) increased, suggesting that oxidative stress was reduced in the liver. Several other findings support this interpretation; the downregulation of 24 ubiquitin proteolytic pathway and related genes indicated a reduced degradation of oxidatively modified or damaged proteins. These findings could be confirmed with the increased nitrogen retention of the birds. Based on the alterations of the polyunsaturated fatty acid (PUFA) biosynthesis genes *FADS1* (fc: -1.22), *ELOVL5* (fc: -1.23), and *ACSL5* (fc: -1.75) the fatty acid (FA) profile was measured in the liver. The PUFA concentration was increased by 13.8% in the Se-yeast group. The literature review revealed that the expression of the PUFA biosynthesis enzymes was changed, due to the altered FA profile. Therefore it was concluded that the reactive oxygen species (ROS) sensitive PUFA were decomposed to a lesser extent in the Se-yeast group, due to a better protection. ROS challenged PUFA decompose to form reactive lipid aldehydes, which are glutathionylated and decomposed by AKR1B1 (fc: -3.06). The dowregulation of the biomarker for oxidized PUFA was confirmed with the reduced hepatic AKR1B1 activity (-18.5%) in the Se-yeast group. The downregulation of 17 vesicular transport related genes indicates that the membrane protection from ROS challenge was increased in the Se-yeast group. The vesicular transport depends on intact membranes. Changes in the biosynthesis of selenoamino acids related genes indicate the unequal use of L-selenomethionine and L-selenocysteine, which were present in the feed supplement at 83% and 5%, respectively.

The expression analysis in the liver of broilers showed that overall effects of a supplement can be measured and the magnitude of the effects on the cells of an organ can be evaluated.

In the second experiment naturally DON-contaminated wheat at different doses was fed one-day old broilers. DON is a secondary metabolite of fusarium species and is frequently present on cereals. About 25% of the world's food crops are contaminated with mycotoxins. Even though DON is not as toxic as other mycotoxins, it is one of the most common trichothecenes. Acute and high doses of DON can lead to diarrhea, vomiting, leukocytosis, and hemorrhages. The chronic exposure can reduce feed intake, weight gain, and stimulate or suppress the immune system, depending on frequency and dose. On the molecular level DON binds to the 28S ribosomal RNA and causes its cleavage, which leads to the inhibition of protein synthesis. Organs with a high protein turnover rate, such as the liver, are therefore more severly affected. In the animal trial uncontaminated wheat was exchanged with naturally DON-contaminated wheat up to the intended concentration of 1, 2.5, and 5 mg/kg DON (groups: DON1, DON2.5, DON5). The four groups with 5 broilers each were kept for 23 days. The gene expression analysis of 3 broiler livers per group resulted in 367 significantly upregulated and 199 significantly downregulated genes. An influence of DON was observed on the nutrient uptake gene *SLC27A4* (fc: +1.43, DON5), which is a palmitate transporter as well as on the D-serine and neutral amino acid transporter *SLC7A10* (fc: 1.48, DON5). The fructose and glucose transport seemed to be influenced by *SLC2A5* (fc: -1.54, DON2.5). The downregulation of the translation initiation regulatory genes *EIF2AK3* (fc: -1.29, DON2.5/5) and *DNAJC3* (fc: -1.44, DON2.5) seemed to counterbalance the 28S ribosomal RNA inhibition of DON. The proteins PARP1 (fc: 1.41, DON5), MPG (fc: 1.41, DON5), EME1 (fc: 1.37, DON5), XPAC (fc: 1.28, DON5) and CHAF1B (fc: 1.44, DON5) are related to single stranded DNA repair. The upregulation of those genes indicate that the DNA is modified at the dose of 5mg/kg DON. This finding might be connected to the downregulation of *AKR1B1* (fc: -3.51, DON2.5/5), which detoxifies mutagenic, reactive lipid aldehydes. The proteins EXOSC9 (fc: -1.50, DON5) and XRN1 (fc: -1.80 DON5) are related to the degradation of specific mRNA's. The deregulation of the tight junction protein *TJP1* (fc: -1.52, DON5) and *CLDN3* (fc: 1.64, DON1/5) seems to be related to the observed decrease of the transepithelial electrical resistance (TEER) in cell cultures upon DON administration. The maximal allowed DON concentration for poultry feed in the European Union is 5mg/kg DON. In our study already at 2.5mg/kg DON significantly altered gene expression was observed.

In the third experiment, twenty one-day old broilers per group were fed the following diets for 35 days: control feed, feed with DON-contaminated wheat (D20), and feed with DON-contaminated wheat and 0.2% polymeric glucomannan mycotoxin absorber (Mycosorb®, Alltech Inc., KY, USA) (MD20). The measured DON concentration in D20 and MD20 was 17.7mg/kg feed. Gene expression was determined in the livers of seven animals from the control and group MD20 each, and eight animals from group D20.

The body weight in D20 was significantly reduced at the end of week two and three; however the mycotoxin-absorber prevented the birds from MD20 from the growth depression. Until the end of the trial the birds from D20 recovered and the body weight was similar in all groups. The feed consumption was significantly reduced in D20 in week three and over the entire feeding period. The total feed consumption was also significantly reduced in MD20, although to a lesser extent than in D20. The gene expression analysis resulted in 195 significantly regulated genes in D20, 433 genes in MD20 and 635 in the comparison between D20 and MD20. Only 1.5% of all regulated genes were altered in both D20 and MD20, and 63.1% were regulated between D20 and MD20. This shows that the functional mechanism is completely different in D20 compared to MD20.

In D20 genes were mainly related to the xenobiotic metabolism and to the prevention of the cell cycle arrest in the G_2/M phase. Beside the detection of the cell cycle transcription factors E2F1, E2F2, E2F4 and E2F6, the expression change of *CDC2* (fc: 1.50, D20) seems to have a major influence on the G_2/M transition. An indication for the delayed development of the birds is also the increased apparent metabolizable energy (AME) of D20 in the last week. The AME in D20 was similar to the first measurement (day 15-17) in contrast to the decreased AME in control and MD20. In MD20 the monolayer permeability seems to be influenced by the gap and tight junction proteins GJA1 (fc:-1.52), GJC1 (fc: -1.92) and CLDN2 (fc: -2.09). The melanocortin receptor MCR5 (fc: 1.64, MD20) and the serotonin biosynthesis enzymes DDC, GCH1 and GCHR seem to influence the hormone signaling (α-melanocyte-stimulating hormone, serotonin). The low-density lipoprotein (LDL) receptor (LDLR, fc: -1.70, MD20, D20 vs. MD20) and the LDL-shell protein APOB (fc: 1.24, MD20, D20 vs MD20) might have caused the significant increase of LDL-cholesterol by 22.2% in heparinized blood plasma of the MD20 group. Totally 13 of 23 genes responsible for cholesterol biosynthesis were downregulated in MD20 even though the total cholesterol was not altered and the greatest concentration decrease was observed from D20 to MD20. The suppression of the immune system seems to be influenced by the *suppressor of cytokine signaling 2* (*SOCS2*, fc: -1.48, D20, fc: -1.72 MD20), *toll-like receptor 5* (*TLR5*, fc: -1.64, D20 vs. MD20) and *interleukin 1B* (*IL1B*, fc: -1.37, MD20). TLR5 recognizes pathogen-associated molecular patterns, and its downregulation together with the downstream cytokine IL1B might be responsible for a reduced reaction upon the challenge with pathogens. The downregulation of the 28S RNA cleavage enzyme RNASEL (fc: -1.41, MD20) seems to be prevent the downstream effects of the DON binding to the 28S RNA. In the comparison D20 vs. MD20 the deregulation of the insulin signaling pathway led to the increased expression of the gluconeogenesis gene *PCK1* (fc: 7.76, MD20, D20 vs. MD20). The altered insulin signaling seems to be caused over the insulin receptor (fc: 1.52, D20 vs. MD20), insulin receptor substrate (GAB1, fc: 1.20, D20 vs. MD20), PIK3R1 (fc: -2.12, MD20, D20 vs. MD20), AKT and the PCK1-expression regulators FOXO1 (transcription factor analysis) and PPARGC1A (+2.49, MD20, D20 vs. MD20).

The expression analysis and performance parameters showed that the mycotoxin-absorber could protect the birds from detrimental effects. However a high number of genes with an altered expression in MD20 could be related to effects that are specific for DON.

Zusammenfassung

Die Untersuchung zu den Effekten von Futtermittelzusatzstoffe und –kontaminanten auf der Ebene von Metaboliten, Proteinaktivitäten und –mengen ist sehr zeitaufwendig und setzt ein umfassendes Wissen über mögliche Abläufe in der Zelle voraus. Aus diesem Grund wurde die Genexpression in der Leber von Broilern gemessen. Dadurch kann das ganze Transkriptom nach Genen mit unterschielicher Expression abgesucht werden. Die Leber wurde für die Genexpression ausgewählt, weil sie das zentrale Organ für den Metabolismus und die Entgiftung ist. Zudem gelangen alle in die Pfortader resorbierten Nährstoffe zunächst zur Leber, die einen Teil davon aufnimmt und umbaut.

Für die Genexpressionsanalyse wurden Chicken Genome Arrays (Affymetrix, Santa Clara, CA, USA) verwendet, welche 32'000 verschiedene Testsequenzen enthalten, die für ungefähr 28'000 Transkripte kodieren. Die Expressionsänderung zwischen den Gruppen musste mindestens das 1.2-fache (fc) betragen und der Signfikanzwert musste kleiner als 0.05 sein.

In drei Projekten wurde die Genexpression in der Leber von Boiler bestimmt, abhängig von der Supplementierung des Futters mit Selen und unterschiedlichen Konzentrationen von Deoxynivalenol.

Für die Evaluation und die Etablierung der Genexpressionstechnik wurden den Tieren im ersten Projekt selenierte Hefen (Se-yeast) gefüttert. Das Selen in Se-yeast liegt hauptsächlich in Form von L-Selenomethionin vor, das auch unspezifisch in Proteine eingebaut werden kann. Selen ist ein essentielles Spurenelement und ein Kofaktor der antioxidativen Enzyme Glutathion Peroxidasen (GPX). Im Tierversuch wurde 16 Broiler pro Gruppe entweder das Kontrollfutter oder ein mit 0.5 mg/kg Selen supplementiertes Futter für 35 Tage gefüttert.

Insgesamt wiesen in der Se-yeast Gruppe 104 Gene, im Vergleich zur Kontrolle, eine signifikant erhöhte und 349 Gene eine signifikant niedrigere Genexpression auf. Der Selenstatus der Tiere von der Se-yeast Gruppe war erhöht, aufgrund der 2-fach höheren GPX-Aktivität in heparinisiertem Blutplasma. Neben der erhöhten Aktivität, war auch die Expression von *GPX1* (fc: 4.16) und *GPX4* (fc: 1.43) erhöht, was ein Hinweis auf reduzierten oxidativen Stress in der Leber ist. Zudem wurden weitere Anzeichen für einen reduzierten oxidativen Stress in der Leber gefunden; die Herabregulierung von 24 Genen der Ubiquitin Proteolyse ist ein Hinweis, dass der Abbau von oxidativ modifizierten und defekten Proteinen reduziert war. Dieser Befund konnte mit der erhöhten Stickstoffretention der Tiere bestätigt werden. Aufgrund der veränderten Expression der mehrfach ungesättigten Fettsäuren (PUFA)-Biosynthesegene *FADS1* (fc: -1.22), *ELOVL5* (fc: -1.23), und *ACSL5* (fc: -1.75) wurde das Fettsäureprofil in der Leber gemessen. Die PUFA Konzentration war in der Se-yeast Gruppe 13.8% erhöht. Anhand der Literaturrecherche wurde klar, dass die Expressionänderung der PUFA Biosynthesegene durch das veränderte Fettsäuremuster ausgelöst wurden. Aus diesem Grund wurde die Schlussfolgerung gezogen, dass PUFA, welche sehr anfällig für oxidativen Stress sind, in der Se-yeast Gruppe in geringerem Masse verbraucht werden. Oxidierte PUFA zerfallen zu reaktiven Lipidaldehyden, die wiederum glutathionyliert und anschliessend von AKR1B1 (fc: -3.06) abgebaut werden. Die Herabregulierung des Biomarkers für oxidierte PUFA konnte mit der reduzierten, hepatischen AKR1B1 Aktivität (-18.5%) in der Se-

yeast Gruppe bestätigt werden. Die Herabregulierung von 17 Genen des vesikulären Transports zeigt auf, dass der Membranschutz gegen reaktive Sauerstoffarten in der Se-yeast Gruppe verbessert ist. Die veränderte Expression von Selenoaminosäure-Biosynthese verwandten Genen zeigt den unterschiedlichen Verbrauch von L-Selenomthionin und L-Selenocystein auf, welche zu 83% respektive 5% in den Selenhefen vorhanden sind.
Die Geneexpressionsanalyse zeigte, dass die gesamten auftretenden Effekte eines Supplements gemessen werden können und deren Ausmass auf die Zellen eines Organs evaluiert werden kann.

Im zweiten Experiment wurde natürlich Deoxynivalenol (DON)-kontaminierter Weizen in unterschiedlichen Dosen an einen Tag alte Broiler verfüttert. DON ist ein sekundärer Metabolit, produziert von *Fusarium*-Arten, welcher regelmässig in Getreide wie Weizen und Mais, vorkommt. Ungefähr 25% des weltweiten Getreideertrags sind mit Mykotoxinen kontaminiert. Obschon DON eine tiefere Toxizität als ander Mykotoxine aufweist, tritt es am häufigsten auf. Eine akute, hohe Dosis DON kann zu Durchfall, Erbrechen, Leukozytose und Blutungen führen. Die chronische Belastung mit DON führt zu reduzierter Futteraufnahme, Gewichtszunahme und stimuliert oder unterdrückt das Immunsystem, abhängig von der Häufigkeit und Konzentration von DON im Futter. Auf der molekularen Ebene bindet DON an die 28S ribosomale RNA und verursacht deren Spaltung. Dies wiederum führt zur Inhibition der Proteinsynthese. Deshalb sind Organe mit einem hohen Proteinumsatz, wie die Leber, am stärksten beeinträchtigt.
Im ersten Versuch wurde unkontaminierter Weizen mit natürlich DON-kontaminiertem Weizen ausgetauscht, um die beabsichtigten DON-Konzentrationen von 1, 2.5 and 5 mg/kg zu erreichen (Gruppen: DON1, DON2.5, DON5). Die vier Gruppen von je fünf Broilern wurden für 23 Tage gehalten. Die Genexpression in der Leber von drei Broilern pro Gruppe ergab 367 signifikant heraufregulierte Gene und 199 signifikant herunterregulierte Gene. DON scheint die Aufnahme von Nährstoffen aufgrund der veränderten Expression von *SLC27A4* (fc: +1.43, DON5), einem Palmitinsäurtetransporter und *SLC7A10* (fc: 1.48, DON5) einem Transporter von D-Serine und neutralen Aminosäuren zu beeinflussen. Die Herabregulierung der Translationsinitiationsregulatoren *EIF2AK3* (fc: -1.29, DON2.5/5) und *DNAJC3* (fc: -1.44, DON2.5) scheint die Inhibition der 28S ribosomalen RNA zu kompensieren. Die Proteine PARP1 (fc: 1.41, DON5), MPG (fc: 1.41, DON5), EME1 (fc: 1.37, DON5), XPAC (fc: 1.28, DON5) und CHAF1B (fc: 1.44, DON5) sind in die Einzelstrangreparatur der DNA involviert. Die Heraufregulierung dieser Gene zeigt, dass die DNA bei einer Konzentration von 5mg/kg DON sehr wahrscheinlich modifiziert wird. Dieses Resultat steht vermutlich im Zusammenhang mit der Herabregulierung von *AKR1B1* (fc: -3.51, DON2.5/5), einem Gen, das mutagene, reaktive Lipidaldehyde abbaut. Die Proteine *EXOSC9* (fc: -1.50, DON5) und *XRN1* (fc: -1.80 DON5) sind in den Abbau von spezifischen mRNAs involviert. Die Deregulation der Tight Junction Proteine TJP1 (fc: -1.52, DON5) und CLDN3 (fc: 1.64, DON1/5) scheint mit der reduzierten, transepithelialen elektrischen Resistenz in Zellkulturen, in Verbindung zu stehen, deren Medien mit DON versetzt wurden. Die Maximalkonzentration für Geflügelfutter darf in der Europäischen Union 5 mg/kg

DON betragen. In unserer Studie wurden bereits bei 2.5 mg/kg DON Gene mit einer veränderten Expression gefunden.

Im dritten Versuch wurden an zwanzig Eintagesküken pro Gruppe die folgenden Futtervarianten für 35 Tage gefüttert: Kontrollfutter, Futter mit DON-kontaminiertem Weizen (D20) und Futter mit DON-kontaminiertem Weizen und 0.2% eines Mykotoxinebinders in Form eines Glucomannanpolymer (Mycosorb®, Alltech Inc., KY, USA) (MD20). Die gemessene DON Konzentration in D20 und MD20 betrug 17.7 mg/kg. Die Genexpression wurde in der Leber von je sieben Tieren der Kontrollgruppe und der Gruppe MD20, sowie von acht Tieren der Gruppe D20 gemessen.
Das Körpergewicht in D20 war am Ende der zweiten und der dritten Woche signifikant reduziert. Dieser Effekt konnte vom Mykotoxinabsorber in Gruppe MD20 verhindert werden. Bis zum Ende des Versuchs erholten sich die Tiere der Gruppe D20 und das Körpergewicht war bei beiden Gruppen ähnlich. Die Futteraufnahme war in Gruppe D20 während der dritten Woche sowie über die ganze Mastperiode signifikant reduziert. Der totale Futterverzehr war auch in MD20 über die ganze Mastdauer signifikant herabgesetzt, obschon die Reduktion geringer ausfiel als in D20. Mit Hilfe der Genexpression wurden 195 Gene in D20, 433 Gene in MD20 und 635 Gene im Vergleich D20 vs. MD20 identifiziert, die alle eine Signifikanz aufwiesen. Nur gerade 1.5% aller regulierten Gene waren sowohl in D20 als auch in MD20 zugegen und 63.1% aller Gene waren im Vergleich D20 vs. MD20 reguliert. Dies zeigt auf, dass die Funktionsmechanismen von DON in D20 und MD20 komplett verschieden sind.
Eine veränderte Expression in D20 hatten Gene des xenobiotischen Metabolismus und solche zur Verhinderung des Stillstands im Zellzyklus in der Phase G_2/M. Neben der Detektion der Transkriptionsfaktoren für den Zellzyklus, E2F1, E2F2, E2F4 und E2F6, war die Expression von *CDC2* (fc: 1.50, D20) verändert, was einen wichtigen Einfluss beim Übergang der Phase G_2/M hat. Ein weiterer Hinweis auf die verzögerte Entwicklung ist die erhöhte, scheinbar umsetzbare Energie (AME) in D20, welche in der letzten Woche gemessen wurde. Die AME in D20 während der letzten Woche war ähnlich zu den Messungen in allen Gruppen während der dritten Woche, im Gegensatz zu der tieferen AME in der Kontrolle und MD20. In MD20 scheint die Durchlässigkeit der monomolekularen Schicht durch Gap und Tight Junction Proteine GJA1 (fc:-1.52), GJC1 (fc: -1.92) und CLDN2 (fc: -2.09) beeinflusst zu sein. Der Melanocortin Rezeptor MCR5 (fc: 1.64, MD20) und die Serotonin Biosyntheseenzyme DDC, GCH1 und GCHR scheinen die Hormonsignalkette (α-Melanozyten-stimulierendes Hormon, Serotonin) zu beeinflussen. Der Low-density Lipoprotein (LDL) Rezeptor (LDLR, fc: -1.70, MD20, D20 vs. MD20) und das LDL-Mantel Protein APOB (fc: 1.24, MD20, D20 vs MD20) scheinen die erhöhte Konzentration von LDL-Cholesterol (22.2%, MD20) in heparinisiertem Blutplasma ausgelöst zu haben. Im Gesamten waren 13 von 23 Genen der Cholesterolbiosynthese herabreguliert, obschon die Konzentration des Gesamtcholesterols nicht signifikant verändert war und die grösste Abnahme der Konzentration von D20 zu MD20 beobachtet wurde. Die Unterdrückung des Immunsystems scheint durch die Gene *Suppressor of Cytokine Signaling 2* (fc: -1.48, D20, fc: -1.72 MD20), *toll-like receptor 5* (*TLR5*, fc:

-1.64, D20 vs. MD20) und *Interleukin 1B* (*IL1B*, fc: -1.37, MD20) ausgelöst zu werden. TLR5 erkennt erregerassoziierte molekulare Muster, und dessen Herabregulierung zusammen mit dem nachgeschalteten Cytokin IL1B scheinen verantwortlich für eine reduzierte Reaktion auf den Angriff von Pathogenen zu sein. Herabregulierung des 28S RNA Spaltungsenzyms RNASEL (fc: -1.41, MD20) scheint die nachfolgenden Effekte der Bindung zwischen DON und der 28S RNA zu verhindern. Im Vergleich D20 vs. MD20 führte die Deregulation des Insulinsignalwegs zur erhöhten Expression des Gluconeogenesegens *PCK1* (fc: 7.76, MD20, D20 vs. MD20). Die Deregulation des Insulinsigalwegs scheint verursacht durch den Insulinrezeptor (fc: 1.52, D20 vs. MD20), das Insulinrezeptorsubstrat (GAB1, fc: 1.20, D20 vs. MD20), PIK3R1 (fc: -2.12, MD20, D20 vs. MD20), AKT und den PCK1-Expressionsregulatoren FOXO1 (Transkriptionsfaktoranalyse) und PPARGC1A (+2.49, MD20, D20 vs. MD20).

Die Expressionsanalyse und Leistungsparameter zeigten, dass der Mykotoxin-Absorber die nachteiligen Effekte von DON mindestens teilweise verhindern konnte. Trotzdem konnte die veränderte Expression einer grossen Zahl von Genen in MD20 DON spezifischen Effekten zugeordnet werden.

1. General introduction

The continuous exposure to feed throughout a lifetime renders diet the most important environmental factor challenging the biological system of animals and humans [1, 2]. Nutrition has become a growing concern with the increasing production levels of farm animals and the increased quality recruitments of animal products.

Nutritional strategies and dietary manipulation are key approaches for influencing the performance and the health status of farm animals [1, 2]. The molecular mechanisms and effects influenced by an altered composition of the diet are very difficult to detect in a comprehensive manner. The transcriptomics method to evaluate and characterize a feed supplement offers fertile opportunities for study. The method allows a screening for differently expressed genes and the subsequently detection of molecular effects and influenced pathways in a defined tissue. This method is the subject of this PhD thesis.

In human nutrition, the focus is mainly on health issues and on disease prevention [1, 3]. Two aspects are substantial: first, the knowledge about the molecular mechanisms and effects that are influenced may lead to new approaches for human studies, and second, the quality of the animal products and the concentration of certain ingredients, such as altered concentrations of polyunsaturated fatty acids or cholesterol concentrations, may have effects on chronic disease progression in humans and on general
human health.

Prior to the sequencing of the genome, it was not possible to take an integrative metabolism approach to the exploration of diet effects. Therefore, the majority of the studies were obliged to use common and well-characterized biomarkers (e.g. glutathione peroxidase activity, blood plasma cholesterol and glucose) to improve knowledge about the mode of action of a nutritional component. Even though those studies improved our knowledge, they suffered from a shared inherent quandary - that the study design used current dogma [1]. This led to a reduced ability to identify novel mechanisms of action and biomarkers that would better explain the health status or the performance parameters of an organism. The choice for measured parameters in nutritional studies was based on previous findings of a nutritional component or a related biological mode of action. The coverage of the effects, therefore, was biased and novel effects were often discovered only with high effort and the measurement of a high number of parameters.

The advantage of the transcriptomics approach is the parallel measurement of all transcripts at the same time, in a single sample, and with only a single sample preparation. Especially important is the fact that the sample preparation highly reduces the risk for variability between measurements. In addition the biomarkers used in traditional disease studies were detected at endpoints, as in LDL-cholesterol in cardiovascular diseases, albeit that other biomarkers may have explained more accurately the ongoing of a disease [1, 3].

The use of microarray analysis should lead to a coverage of all effects in an organ at the level of transcription and it holds the potential to uncover the molecular causes of changed performance parameters.

Nutrigenomics

The development of human and animals is defined by environmental influences, like diet, physical activity, pathogen challenge, and heredity, which shows that both aspects must be considered for the improvement of health and animal performance parameters. Even though the interaction of environmental influences and genetic background is well accepted, the focus remains either on genes or on environment, but both have not been analyzed simultaneously. By the completion of the sequencing of organisms like humans, mice, and chickens, it became possible to analyze globally the messenger RNA (mRNA) (transcriptomics) and the genes (genomics). Other techniques have also evolved at the level of proteins (proteomics) and metabolites (metabolomics) [1]. Today, theoretically, the expression of all genes can be measured with transcriptomics in contrast to proteomics and metabolomics, which can measure only a fraction of the proteins and the metabolites, respectively [3]. A derivative of these techniques is functional genomics, which aims to uncover the functional role of different genes and how these genes interact and influence each other. The most widely used field is still transcriptomics, which analyzes the gene expression (mRNA) of a biological sample to a certain time under a specific condition [1].
Nutrigenomics aims to identify and understand the molecular mechanisms between nutrients and the genome or the gene expression, respectively, with the use of microarrays and bioinformatics. Microarray analysis is a quantitative assessment of the relative tissue concentration of the specific mRNA, which is directly related to the expression of a gene (Fig. 1.1) [4]. First, the total RNA is extracted from the target tissue and cleaned up, to remove DNA and other contaminants in the sample. Then, a reverse transcription, using T7 oligo (dT) promoter primer, is performed, followed by a second strand synthesis to produce cDNA, which corresponds to the sequence of the mRNA. The cDNA is then used as a template for the in vitro transcription (IVT) and the simultaneous labeling of the cRNA. The cRNA is subsequently fragmented and hybridized on the microarray. The intensity of the fluorescence from the hybridized cRNA with complementary probe sets is directly related to the amount of the target mRNA in the source tissue and reflects the expression of a particular gene. This allows the determination that a gene is up- or down-regulated compared to the control samples, as a consequence of a specific biological manipulation or a dietary treatment [4]. The methods for the analysis of the results from the intensity scan of the microarrays have evolved in the past. Therefore, the Functional Genomic Data Society has developed guidelines (minimum information about a microarray experiment [MIAME]) that specify the minimum standards necessary to ensure proper interpretation of microarray results [5].

Figure 1.1: Workflow for the gene expression analysis

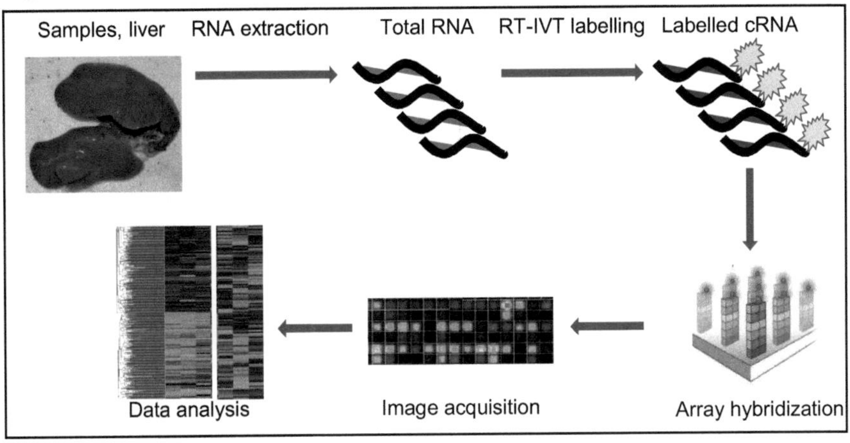

The schematic diagram shows the workflow of the gene expression analysis. First the total RNA is extracted from tissues or cell cultures. Then the mRNA is reversely transcribed into cDNA and subsequently an *in vitro* transcription is performed to produce labeled cRNA. For the hybridization to the probes on the microarray the cRNA copies are fragmented. The intensity of the signal on the microarray corresponds to the expression of a gene in the origin sample. The intensity values are then normalized and the expression level of each gene is compared between treatments and significance of differences is determined.

It should be cautioned that an altered expression of a gene does not lead compellingly to an alteration of the protein amount or the related metabolites. But the altered expression of a gene, and gene clusters with a biological relationship, are often strong indicators for changes in homeostasis and metabolism. Therefore, verification on the level of protein amount, enzyme activity, and the amount of metabolites is obligatory. Under the condition that a parameter is significantly changed, the microarray analysis has already resulted in the identification of the candidate gene or gene cluster. The knowledge about the involvement of a specific gene or cluster subsequently allows researchers to identify pathways causing transcription factors, signal transducers, degradation proteins and so on, which are related to the observed gene expression [1, 3].

Nutrition and gene expression

Transcriptomics is widely used in pharmaceutical and nutritional research. Whereas pharmaceuticals have a targeted approach with the "one drug one target" paradigm, diet has a multi-parametric approach, because it is comprised of a multitude of nutritional and chemical molecules, each capable of regulating disparate biological processes [1]. In relation to human nutrition, food-intake is a key environmental factor, which affects the incidence of chronic diseases like obesity, diabetes, cardiovascular diseases, and cancer [1, 3, 4]. The exact number of biologically active components influencing chronic diseases is unknown. Therefore, the understanding of how feed and food components can modulate homeostasis may help to improve prevention and management of nutrition-induced, chronic diseases. Nutrients are also considered as "signaling molecules"[4], which lead through cellular sensing mechanisms to changes in gene and protein expression, mainly over transcription factors with subsequent effects on the concentration of metabolites. This approach allows explanations of the mechanisms at the molecular level, when a certain diet is consumed.

The molecular structure of a nutrient is mainly responsible to activate specific pathways, unsaturated fatty acids or plant sterols, for example, in contrast to saturated fatty acids or cholesterol, respectively [3]. One of the major challenges is to characterize the influenced pathways and the downstream effects of a nutrient. The use of genome-wide measurement of nutrient target genes should allow explanations of how nutrients act and how they influence health and disease. In addition, nutrigenomics should lead to the development of evidence-based, healthful feed and food, as well as optimized nutrient content in regard to the performance of farm animals [3]. The aim, in general, is to find new biomarkers for a diet, new bioactive food ingredients, and the validation of bioactive ingredients. In human nutrition the concept of "personalized diets" is pursued, to account for the different genotypes, which join nutrition in influencing the progress of chronic and complex diseases, like cardiovascular disease, insulin resistance, or diabetes [3]. Depending on the genotype, an ingredient of a diet could have an influence on the health status of an individual. Thus, specific genetic information is needed for the definition of an optimal diet for an individual or group of animals with a high genetical comparability between individuals [6].

Nutrition and animal models

In general animal nutrition research focuses often on applied, mission-orientated projects and there seems to be a segregation between agricultural research and human medicine, veterinary medicine and the basic life science disciplines [7]. Therefore the use of molecular tools in animal nutrition might improve the comparability to human studies and the animal models rats and mice, which are the overwhelming models of choice for basic research [7]. The mouse model has several advantages - the small size and the relatively low cost of maintenance. Farm animals are bigger but surgery, blood sampling, tissue recovery, and many other biomedical applications are more easily achieved with larger animals. For certain human diseases, it has been shown that farm animals are interesting models and that disease progressions are highly similar to those observed in humans. The baby pig is often used as a model for human infant nutrition and pigs develop chronic diseases like atherosclerosis, cardiovascular disease, and obesity [7, 8]. According to Roberts *et al.* (2009), seventeen Nobel Prize winners have used farm animals, like cattle, pigs, sheep, goats, horses and chickens, as research models [7].

The microarray analysis technique should be established and evaluated with the use of broiler chickens at the Institute of Agricultural Sciences, ETHZ. The knowledge gained from this research and thesis should help to get the fundament for the use of farm animals as models for animal and human disease studies with nutritional genomics; it should also increase the analysis possibilities at the Institute. The products from the farm animals are mainly consumed by humans and those products influence their health in general. It has been shown that diets with a balanced omega-3/omega-6 ratio can improve health and can contribute to the prevention of multiple chronic diseases [6]. Thus, the use of nutrigenomics should also identify new approaches to improve animal products, which are dedicated to human consumption.

The modification of the gene expression in a tissue of a living animal is also a good opportunity for the evaluation of the function of a target gene and its up- or down-stream effects in basic research. The advantage of nutrigenomics is that a deregulation of the gene expression occurs under conditions, which are often common, like the supplementation of selenium, or which might occur infrequently, like the contamination with deoxynivalenol. Neither influence intervenes, technically, into the gene expression, like knock-down or -out technologies, which might have unintended influences on other pathways, especially the feedback loops, which might be activated in knockdown or -out studies, leading to other effects compared to expression changes caused by a nutrient [9]. A decreased transcription of a gene is caused by an altered expression regulation. Meanwhile, in knockdown or -out studies, the expression change is independent of the transcription regulation. Therefore, the addition of a supplement to a diet, with the aim to alter the expression of a target gene, has the potential for basic research and the study of molecular interactions.

Projects

Three animal experiments were performed in this PhD thesis. For the evaluation and implementation of animal nutrigenomics, selenium was used as a feed additive. The effects of selenium are well characterized

[10], which allows a comprehensive comparison between the results in this PhD thesis and the literature. In the second and third experiment, deoxynivalenol (DON), a frequently detected mycotoxin in contaminated feedstuff, was added to the diet of the broilers in different doses. The knowledge about the molecular effects of DON has been elucidated in recent years in cell cultures; however, its effects on living animals remain to be evaluated [11-13]. This was the reason to explore the effects in the liver of broilers to compare, on one side, the results from the *in vitro* studies to the results in the living animal. On the other side, new effects should be detected, with the aim to fund the knowledge for possible strategies against the detrimental effects of DON.

The liver was used in all three experiments, due to its central role in metabolism and in detoxification.

A nutrigenomics approach was chosen for the evaluation of the selenium supplementation and the addition of DON contaminated wheat in the feeding experiment:

- The total RNA was extracted from the livers of birds with an average weight and the gene expression was measured with micorarrays.
- The "Marker" genes and gene clusters, which to have a fold change of at least 1.2 and a significance value below 0.05 compared to the control group, were detected with gene expression analysis,.
- The altered expression of the "marker" genes were verified with reverse transcription real-time PCR.

- The influenced key pathways were identified with the "marker" genes and the gene expression analysis.
- Based on the identified pathways, selected enzyme activities, and metabolite concentrations were measured, and together with the performance parameters, compared to gene expression analysis.
- With the help of a literature review, the biological relevance of the differently expressed genes and gene clusters were evaluated.

Aims and scope

The following objectives should be attained in this PhD thesis:

- The technique of microarray analysis in differently fed broilers should be evaluated, to estimate the advantages, disadvantages and the usefulness of this approach in nutrition research.
- The microarray analysis in the liver of differently fed broilers should be implemented at the ETH Zurich, Institute of Agricultural Sciences, together with the Functional Genomics Center Zurich.
- The results of the gene expression analysis should be verified with real-time PCR, enzyme activity assays, measurements of metabolites and the measurements of performance parameters.
- The effects of supplements or contaminants on the gene expression should be verified on its biological credibility by a literature review.
- The feed supplement selenium in the form of selenized yeast, DON-contaminated wheat and the mycotoxin-absorber Mycosorb should be evaluated with the gene expression analysis in the liver of broilers.

2. Nutrigenomics: Gene expression in broiler livers upon supplementation with selenized yeast

Abstract

Selenium (Se) is an essential cofactor of the antioxidant enzyme glutathione peroxidase (GPX) beside other functions. The evaluation of optimal selenium supplementation in broiler feed and the subsequent effects on animal health and performance requires comprehensive knowledge of the overall metabolic effects of selenium. Gene expression profiling using microarrays should help to gain insight into the effects of supplementation with selenized yeast (Se-yeast), which mainly consists of organic Se, on metabolism-related genes. Therefore the gene expression profile was determined in the livers of four control and five treatment (0.5 mg Se/kg feed) 35-day-old broilers using chicken genome arrays covering more than 32,000 transcripts. Fatty acid (FA) analysis, aldo-keto reductase family 1, member B1 (AKR1B1) activity in the liver, and GPX activity in heparinized blood were determined. Several selected gene expressions were also verified using quantitative PCR (qPCR).

Analysis of the gene expression in the liver using microarrays yielded 110 upregulated and 366 downregulated genes with significant alteration and a minimal fold change of 1.2 by Se-yeast supplementation ($P < 0.05$). The selenium status was determined by a two-fold increase of glutathione peroxidase activity. The following clusters of significantly regulated genes were found: selenoproteins and genes related to selenoamino acid biosynthesis, glutathione metabolism, the ubiquitin cycle, protein degradation, reactive oxygen species regulation, lipid metabolism, and vesicular transport. Subsequently, qPCR confirmed the alterations of the downregulated genes *AKR1B1, FADS1, GSTA1, PDZK1, SEPSECS,* and *TPMT,* and the upregulated genes *GPX1, PDZK1IP1, MRPS18A,* and *SEPP.* Fatty acid analysis of the liver showed significant concentration increases of polyunsaturated fatty acids. The apparent metabolizable energy and nitrogen retention increased significantly in the third week of broiler growth.

The reduced oxidative stress in liver cells, due to the increase in expression of the antioxidant enzymes GPX1 and GPX4, seems to result in decreased PUFA and protein oxidation or damage, respectively. For this reason, protein ubiquitination as the preliminary stage of protein degradation and vesicular transport related to protein biosynthesis, modification, and distribution seems to be reduced. The findings on the level of gene expression are strengthened by the increased N retention and PUFA concentration in the liver.

Indroduction

The essential trace element selenium (Se) is of fundamental importance in animal and human nutrition. It enters the feed and food chain mainly through plants, and therefore the low Se content in European soil can lead to a Se deficiency of humans and animals. In her study, Rayman [10] reviewed feed and food literature and discovered a high variability in Se content in products from different batches, regions, and countries. Another Se food source for humans derives from animal products, such as meat from broilers, which often receive Se-supplemented feed. The enrichment of animal products with Se via feed protects the population against accidental overdoses. Such supplementation seems to be beneficial as low Se levels have been linked to higher mortality, reduced immune response, an increased risk of liver and other cancers, Keshan disease, the myxedematous form of cretinism, atherosclerosis, and reduced fertility [10].

Reactive oxygen species (ROS) are produced in the respiratory chain of mitochondria by cellular respiration. At least 25 selenoproteins have Se as an integral element, which protects cells against ROS and other free radicals [14]. The efficiency of the antioxidant system has major effects on the health and performance of animals. Low levels of Se result in oxidative stress that in turn can harm cell elements, such as lipids, proteins, and DNA due to imbalances of ROS and ROS-scavenging molecules. In living cells, Se regulates glutathione (GSH) and the major Se-dependant antioxidant enzymes, such as glutathione peroxidase (GPX) [15]. Glutathione peroxidase reduces hydrogen peroxide and lipid and phospholipid hydroperoxides and dampens the distribution of free radicals and ROS that are formed during metabolism. Serum selenium concentrations seem to be inversely correlated with *in vivo* markers of oxidative stress, both in humans and in animal models [16]. Glutathione peroxidase activity has been described as a reliable indicator of selenium status in animal and human research [17] and was used to confirm the Se status of the broilers in our study.

Depending on the form of Se supplements, the mode of action is different. For example, the inorganic Se source selenite can be used effectively for selenoprotein synthesis, but L-selenomethionine (SeMet) acts better as storage for selenium. SeMet is incorporated non-specifically into proteins during synthesis; therefore it remains longer in the body and can be stored in tissues [18]. According to Rayman [18], selenized yeast (Se-yeast) for human supplementation, which consists mainly of SeMet, is no longer allowed to be sold in the European Union. The European Community Scientific Committee on Food concluded that Se-yeast supplements are poorly characterized and could potentially cause the buildup of Se in tissues to toxic levels. A broad study of biochemical interactions and effects and analyses of all expressed genes is necessary in order to evaluate this issue. Thus, microarrays that contained 32'000 probe sets representing more than 28'000 transcripts were used. Microarrays are used to cover all possible mechanisms upon Se-yeast supplementation by screening for altered gene expression over the entire transcriptome.

The broiler liver was chosen for the expression analysis because of its central importance in metabolism. In our study, the difference in selenium concentration between the control and treatment groups was 0.5 ppm. The recommended selenium concentration in the broiler diet is 0.15 mg Se / kg (National Research Council, 1994), twice as high as in our control group with 0.07 mg Se /kg. The Se was added as organic Se in the form of Se-yeast, which is characterized by its high bio-availability in different animal species. It consists mainly of SeMet, and to a lesser extent of L-selenocysteine. Selenized yeast is commonly used as a feed additive in animal nutrition. Polyunsaturated fatty acids (PUFA) are among the most susceptible targets of ROS and are

responsible for the integrity and function of membranes. Additionally, C20 fatty acids serve as important precursors for the formation of eicosanoids, such as prostaglandins, thromoboxanes, and leukotrienes [19]. Therefore, a second focus of our study was the fatty acid (FA) profile in the liver, in combination with the expression of genes related to the PUFA biosynthesis pathway.
In nutrition research, detecting small effects of a single dietary component is very difficult. On the other hand, the supplementation of a single compound at higher doses might have negative or unrepresentative effects on the performance and health status of animals. The evaluation of gene expression profiles may help to resolve this discrepancy. Various effects of selenium on protein activities and biochemical parameters have been described. We intend to consolidate several aspects in this study within living animals, and the limits of detectable differences with functional genomics are shown. The effects of Se-yeast on the gene expression of broiler livers has been comprehensively studied and new biomarkers, which might be used for optimizing selenium supplementation, are described. We found several gene clusters that could be related to selenium supplementation and reduced oxidative stress.

Material and methods

Experimental design, birds, and diets

Thirty-two one-day-old male broiler chicks were obtained from a commercial hatchery. The birds were weighed at the beginning of the experiment and separated into two treatment groups with the same total bird weight. After a week, each initial group was divided into three subgroups, due to limited cage size. The birds in all three subgroups were distributed according to similar average weight and received the same diet as in the first week. The broilers were held for 35 days. The animal experiment was performed according to the regulations of the cantonal veterinary office in Zurich (Switzerland). The standard diet (table 2.1) was formulated according to the National Research Council (1994) and prepared for both groups using the same process. Celite, a nutritionally inert substance, was added to the feed as an indigestible marker to increase the level of acid-insoluble ash. After one basic feed mixture was prepared, Se-yeast (25 g/100 kg, Sel-Plex®, Alltech, Lexington, Kentucky, USA) was added to the diet of the treatment group. The pelleted feed and water were provided *ad libitum*. The total cage weight and feed intake were monitored weekly. Total water intake per cage was recorded four days a week. Feed samples were taken from the principle components, the control, and the treatment diet before and after pelleting and on days 1, 17, and 31 of the experiment. All feed samples were stored in a 4 °C cooler. Excreta samples were taken from day 15 to day 17 and from day 29 to day 31 from all six cages. After each sampling, they were frozen at -20 °C. AME and N retention were calculated according to Scott and Hall [20]. Dry matter, ash, acid-insoluble ash, gross energy, and N content were determined for all feed and excreta samples. The selenium concentration was determined in an external laboratory, according to the method of Rodriguez *et al.* (1994) [21].
After 35 days, the broilers were individually weighed and slaughtered by cervical dislocation and bleeding. Blood samples were collected in heparinized blood collection tubes (Venoject, Terumo

Europe, Wettingen, Switzerland), cooled, and centrifuged at 1000 g for 20 minutes. The supernatant was then removed and stored at -20 °C. Liver samples were taken immediately after the blood samples. A slice of approximately 1 cm was cut from the middle part of the left liver lobe and frozen in liquid N. The remainder of the left liver lobe was collected and stored for FA analysis.

Table 2.1: Composition and nutrient content

Ingredient	Broiler control (%)	Broiler Se-yeast (%)
Corn	27.5	27.5
Wheat	35	35
Soybean meal	25	25
Potatoe protein	2.5	2.5
Linseed oil	5	5
Salt	0.15	0.15
DL-methionine	0.26	0.26
Limestone grit /	0.9	0.9
Lysin-HCL	0.17	0.17
DCP 38/40	1.2	1.2
Sodium	0.3	0.3
Celite 545	1.52	1.52
Premix[1]	0.5	0.5
Se-yeast (Se	0	0.025
Analysis		
Selenium (µg/kg)	70	620
Dry matter	88.6	88.0
Crude ash	17.5	18.3
crude protein	25.1	25.4
Fat	6.9	7.1
ADF	3.8	3.8
NDF	10.8	10.4
Gross energy	17.4	17.3

[1]Premix: Diet from Aeschbacher *et al.* (2005) [22]. DCP 38/40: Dicalcium phosphate. ADF: Acid detergent fiber. NDF: Neutral detergent fiber.

Activity measurement of GPX and AKR1B1

Glutathione peroxidase activity was measured in 15 control and 14 Se-yeast heparinized blood samples according to the protocols of Sigma-Aldrich (St Louis, MO,USA, [23]) and Zuberbuehler *et al.* (2006) [24]. One control blood sample was lost during slaughter. The decrease of absorbance of β-nicotinamide-adenine dinucleotide phosphate (NADPH) was measured at 340 nm in a photospectrometer for three minutes. The final assay contained 48 mM sodium phosphate, 0.38 mM EDTA, 0.12 mM β-NADPH (reduced form), 0.95 mM sodium azide, 3.2 units of glutathione reductase, 1 mM glutathione 0.02mM DL dithiothreitol, 0.0007% (w/w) hydrogen peroxide, and 1.6% (v/v) of heparin plasma. The measurement was performed twice with the same samples. Glutathione peroxidase from bovine erythrocytes was used for the standard curve. All reagents and enzymes were purchased from Sigma-Aldrich (Buchs, Switzerland).

Aldo-keto reductase family 1, member B1 (AKR1B1) activity was measured in 14 liver samples per group, according to Kaiserova *et al.* (2006) [25]. Each assay was carried out in duplicate. Briefly,

1g of the liver sample was homogenized in 4 ml of 0.01M potassium phosphate (pH 7.0) containing 1mM ethylenediaminetetraacetic acid (EDTA), 50mM N-ethylmaleimide, and protease inhibitor cocktail (Sigma-Aldrich, Buchs, Switzerland). The homogenate was centrifuged at 105'000g for 1 h at 4 °C and the supernatant was desalted on a Sephadex G-25 (PD-10) column (GE Healthcare Europe, Glattbrugg, Switzerland). The PD-10 column was equilibrated with N_2-equilibrated 50 mM potassium phosphate (pH 6.0), containing 1mM EDTA. The protein amount was determined using a Qubit™ Protein Assay Kit (Invitrogen, Basel, Switzerland). Enzyme activity was measured at 25 °C in a 0.6 ml reaction volume, containing 0.3ml homogenate (1.5 mg protein) and 0.1 M potassium phosphate (pH 6.0), 0.15 mM NADPH, and 10 mM glyceraldehyde. The control reaction contained all components, with the exception of glyceraldehyde. The absorbance was measured for four minutes at 340 nm. One unit was defined as the amount of enzyme to oxidize 1 µmol of NADPH/minute/mg protein.

High density oligonucleotide array hybridization

Five animals were selected according to weight from each of the control and treatment groups. The heaviest and lightest five birds were excluded. The samples were homogenized in a mortar filled with liquid N and ribonucleic acid (RNA) was extracted using TRIZOL reagent (Invitrogen, Basel, Switzerland). Next, 95 µg of RNA from each sample was cleaned using an RNeasy Mini Kit (Qiagen, Hilden, Germany); RNA concentration and purity was determined using an ND-1000 photospectrometer (Nanodrop Technologies, Wilmington, DE, USA) and RNA integrity was determined using an Agilent RNA 6000 Nano Kit (Bioanalyzer 2100, Agilent Technologies, Basel, Switzerland). Only samples with an integrity number higher than nine were used. Labeled cRNA probes were generated according to the Affymetrix protocol using a GeneChip® One-Cycle cDNA Synthesis Kit (Affymetrix, Santa Clara, CA, USA). Briefly, the procedure consisted of a reverse transcription of 7 µg of total RNA with a T7 oligo (dT) promoter primer. After second-strand synthesis, cRNA was generated by *in vitro* transcription using a biotinylated nucleotide analog. The cRNA concentration, purity, and integrity were controlled again before the cRNA samples were hybridized on the arrays. Finally, the labeled cRNA was fragmented and hybridized at 45 °C for 16 hours on the chicken genome array (Affymetrix, Santa Clara, CA, USA). For testing purposes, one sample was processed before the others were hybridized on the microarrays. In this sample, all internal quality values were within the standard range, with the exception of the spike in poly-A RNA control oligos. Therefore, this control group sample was excluded from the gene expression analysis. For all other samples, another Poly-RNA control kit was used and the samples corresponded to the quality specifications. The arrays were washed and stained using a GeneChip Fluidics Station and scanned with a GeneChip Scanner 3000, according to the manufacturer's protocol (Affymetrix, Santa Clara, CA, USA). The GeneChip operating software (GCOS, Affymetrix) controlled the scanner, the fluidics station, and the pre-normalization (MAS5.0) of the data.

Reverse transcription and real-time PCR

To verify the results from the microarray experiment and to extend the number of samples, the total RNA was extracted from the remaining liver samples. Experiments for 12 control and 14 Se-yeast samples were conducted in two steps; first, reverse transcription was performed and then real-time polymerase chain reaction (PCR) was conducted with adjusted amounts of cDNA. For the reverse transcription, a High-Capacity cDNA Reverse Transcription Kit (Applied Biosystems, Rotkreuz, Switzerland) was used. The total reaction volume was 20 µl, containing 2 µg of total sample RNA. Oligos were designed for all real-time PCR measurements of two adjoining exons using Primer Express® 3.0 software (Applied Biosystems, Rotkreuz, Switzerland) to exclude possible amplification due to genomic DNA contamination. The probes were labeled with FAM (fluorescein) reporter dye at the 5' end and TAMRA (rhodomine) was used as a quencher at the 3'end. Glyceraldhyde-3-phosphate (GAPDH) was used as an endogenous control to adjust the cDNA concentration. *GPX1* and *AKR1B1* were used to verify the microarray experiment with TaqMan PCR. SYBR® Green was used to detect the genes *APOA5, FADS1, GAL13, GSTA1, MRPS18A, PDZK1, PDZK1IP1, SCLY, SEPP, SEPSECS,* and *TPMT*. The primers (table 2.2) were purchased from Microsynth (Balgach, Switzerland). The 20 µl real-time PCR reaction contained 900 nM of forward and reverse primers, 250 nM of TaqMan® probe, 100 ng of sample cDNA, and 10 µl of TaqMan® Fast Universal PCR Master Mix (2X) (Applied Biosystems, Rotkreuz, Switzerland). The 20 µl SYBR® Green real-time PCR contained 500 nM of primers, approximately 100 ng of sample cDNA, and 10 µl of Fast SYBR® Green Master Mix (Applied Biosystems, Rotkreuz, Switzerland). The amplification was performed in fast mode on the 7500 Fast Real-Time PCR System with SDS Software (Applied Biosystems, Rotkreuz, Switzerland); this was repeated three separate times for TaqMan and four times for SYBR® Green PCR. For both approaches, the cycling program was 20 seconds at 95 °C and then 40 cycles of the following: denaturation for three seconds at 95 °C and annealing and extension for 30 seconds at 60 °C. The results were evaluated using the comparative Ct method ($2^{-\Delta\Delta Ct}$) using SDS Software.

Table 2.2: Oligos for real-time PCR

Gene	Forward primer / Probe	Reverse primer
GAPDH	TTGGCATTGTGGAGGGTCTT	GGGCCATCCACCGTCTTC
Probe	CGTCCATGCCATCACAGCCACAC	
GPX1	CCCTTTCTGGAAGAGCTGGAA	GAGATCCTCCCTGCACTGAAGT
Probe	CGAAGCCACCTCCTGGCCGC	
AKR1B1	TTAAAGAGATTGCAGCCAAGCA	TCACGTTTCTCTGGATGTGGAA
Probe	ACAAAACTGCAGCACAGGTTCTCCTTCG	
APOA5	GGCTTTCCTTCAGATTTGTGAACT	GCCAGCTGACCAGTGACAAG
FADS1	GTCAAGATGAAACGCTATGAAAGG	CCGCGTCATCAGCCACTAC
GAL13	CAGTGGCCATGGTTGTTCCT	CTTTGCCATCGTTGTCATTCTC
GSTA1	AATTTCCCCTCTTGCAGAGTTTT	GCCAGGCTGCAGGAATTTC
MRPS18A	GGGAAGCGGCTGGGTTT	ATTATGGTCGTGTTGCCTTCAGT
PDZK1	CAGAATTCCCACTGTTCTTCACAAT	TCAGGGTTAACGGTGTGTTTGTA
PDZK1IP1	CGATGCTCACCACTGTCTCAA	TTCTGGTGCTGGTGGCAAT
SEPSECS	AGACAAGTTTCAGGAGCGAGAGTT	GCGACATAAAGCCTTTGAAAGTG

SEPP	AATGAGGACTCTGATGGTTGACAA	TGCAGGCTTCCAGATTGGA
SCLY	GCAGCTGAGCAGAATCAAGGA	AAGACCCTTCTTGCCAGTGTTG
TPMT	CCACCGCCTTCCATAAGGA	CCTGCCATTCACCAGAAGGT

Sequence: 5' – 3'

Fatty acid analysis in the liver

Fatty acids were extracted twice from 14 control samples and 12 Se-yeast samples, according to Mondello *et al.* (1978) [26]. Approximately 1 g of the liver sample was homogenized in hexane:isopropanol (3:2, HIP) with the internal standard C19:0 (1mg per sample). The samples were then left for one hour and were subsequently centrifuged. The supernatant was transferred to a new tube and the liquid was evaporated with N gas. Esterification of fatty acid methyl esters (FAME) was performed with 2 ml of methanolic NaOH (0.5 M), which was boiled for three minutes. Next, 3 ml of BF_3 was added and the solution was boiled again for seven minutes. After that, 7 ml of NaCl (0.34 M) and 4 ml of hexane were added. The samples were centrifuged and the supernatant was transferred into an appropriate vial. FAME were separated on an Omegawax™-250 column (30 m × 0.32 mm, d_f 0.25 µm, Supelco®, Sigma-Aldrich, Buchs, Switzerland) after the injection of 1 µl in an HP 5890 gas chromatograph (Agilent Technologies, Basel, Switzerland). Hydrogen was used as the carrier gas at 2.2 ml/minute. The oven was programmed as follows: 160 °C held for 30 seconds, then heated to 190 °C with a rate of 20 °C/minute, heated to 230 °C with a rate of 7 °C/minute, held for five minutes and 20 seconds and heated to 250 °C with 20 °C /minute and held for six minutes. The flame-ionization detector was maintained at 270 °C. As a standard reference, the Supelco® 37 Component FAME Mix (Sigma-Aldrich, Buchs, Switzerland) was used.

Statistical analysis

The alterations in the gene expression profiles between the control and treatment groups were analyzed using GeneSpring 7.3 software (Agilent Technologies, Basel, Switzerland). Data analysis was performed on nine microarrays (four controls, five with additional selenium). Normalization was carried out in the following way: values below 0.01 were set to 0.01, each measurement was divided by the 50th percentile of all measurements in that sample, and each gene was divided by the median of its measurements in all samples. The genes had to be present or marginal and had to have a raw value above the 50th percentile in three of four samples in the control group and in four of five samples in the Se-yeast group. The gene lists from the control and treatment groups were merged together and an ANOVA was performed. The final gene list was separated into groups of downregulated or upregulated genes with a fold change of at least 1.2. All other data was analyzed using Student's t-tests and ANOVAs using the statistical package R (www.r-project.org) to determine significance ($P < 0.05$).

Results

Diet characterization and performance of broilers

Table 2.1 summarizes the results of the analysis of dry matter, ash, crude protein, and gross energy of the broiler feed. All results were close to standardized diets (National Research Council, 1994) or within the range of one standard deviation. Performance and feed consumption of the broilers did not differ significantly between groups (table 2.3); however, body weight (BW) and total feed consumption (FC) were lower in the Se-yeast group (BW day 35: 2215 g, FC: 3319 g) compared to the control group (BW day 35: 2313.0 g, FC: 3447.3 g). The feed conversion rate (FCR) was very similar, with 1.49 g feed/g gain for the control group and 1.50 g feed/g gain for the Se-yeast group. In the first sampling period the apparent metabolizable energy (AME) and nitrogen (N) retention were significantly increased in the Se-yeast group (AME: 15.0, N retention 64.7%) compared to the control group (AME: 14.5, N retention 59.3%). Water consumption did not differ significantly between the groups. In the Se-yeast group, two broilers died during the experiment, one due to a heart attack and one was slaughtered due to leg problems. All other broilers were in good health.

Table 2.3: Performance parameters

	Control	s.d.	Se-yeast	s.d	P-value
BW day 35 [g]	2313	27	2215	85	NS
Total FI [g]	3447	62	3319	171	NS
FCR [g/g)]	1.49	0.01	1.50	0.04	NS
AME P1 [MJ/kg]	11.7	0.21	12.2	0.16	P < 0.05
AME P2 [MJ/kg]	11.1	0.27	11.3	0.05	NS
N Retention P1	59.31	3.22	64.7	1.73	P < 0.05
N Retention P2	55.52	2.98	59.45	2.26	NS

BW: Body weight. FI: Feed intake. FCR: Feed conversion rate (g gain / g feed). AME P1/2: Apparent metabolizable energy in MJ/kg dry matter. P1/2: Period 1 or 2.

Activity measurement of GPX and AKR1B1

The *GPX* activity in the heparinized blood was 2.03 times higher in the Se-yeast group compared to the control group ($P < 0.0001$) (table 2.4). The AKR1B1 activity was significantly reduced by 18.5% in the Se-yeast group compared to the control.

Table 2.4: Gluthation peroxidase and AKR1B1 activity

		Control		Se-yeast		
		Average (U/µl)	s.d.	Average (U/µl)	s.d.	P-value
GPX	Blood	1.10	0.01	2.23	0.07	$P < 10^{-5}$
AKR1B1	Liver	3.51	0.99	2.86	0.83	P < 0.05

s.d.: Standard deviation.
GPX activity based on GPX activity from bovine erythrocyte.
AKR1B1 activity is the amount of enzyme used to oxidize 1µmol of NADPH/minute

High density oligonucleotide array hybridization

After total RNA extraction, the A260/A280 ratio was between 1.91 and 2.01, the rRNA ratio (28s/18s) was between 1.4 and 1.6, and the RNA integrity number was at 9.2 or higher and showed undegraded RNA samples as analyzed by the Bioanalyzer 2100 (Agilent Technologies, Basel, Switzerland). All the microarray raw data have been deposited into the Gene Expression Omnibus (National Center for Biotechnology Information, http://www.ncbi.nlm.nih.gov/geo) as accession number GSE25151. After normalization and filtering, it was found that 475 probe sets corresponding to 453 genes were significantly regulated in the treatment group compared to the control group (figure 2.1). In the upregulated gene list with 22.9% of the regulated genes, 29.6% of the genes had an unknown function and 4.8% of the genes had a fold change (fc) higher than two. The entire list of 110 upregulated and 365 downregulated probe sets is available in the section Appendix A. Table 7.1 " Selenium experiment: Upregulated genes" (Appendix A) contains the discussed 104 upregulated genes with a minimal fc of 1.2 and a maximum P value of 0.05.
The 349 downregulated genes in the Se-yeast group are listed in table 7.2 " Selenium experiment: Downregulated genes" (Appendix A). Out of the original list with all regulated genes, 77.1% of the genes were downregulated. In the group of downregulated genes, 31.1% of the genes were not annotated and 2.0% of the genes showed more than a two-fold change.

Figure 2.1: Heat map and Pearson correlation of significantly altered genes.

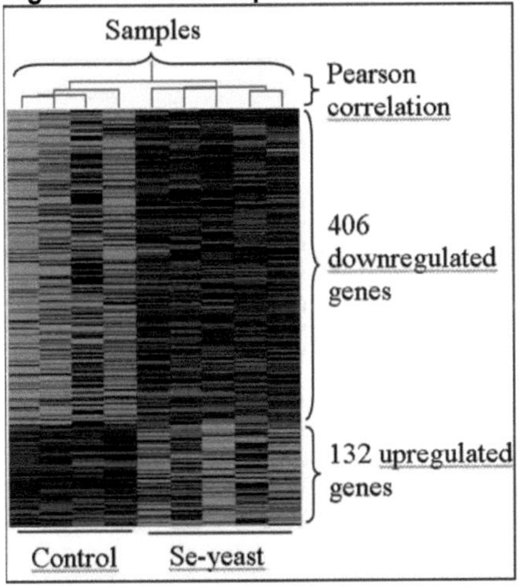

The heat map and the Pearson correlation were generated with all 544 significantly altered gene expressions. The genes with a higher expression are indicated in green compared to the opposite group and the genes with a lower expression are indicated in red. The Pearson correlation was calculated with GeneSpring 7.3 (Agilent Technologies, Basel, Switzerland) and shows the relative distance between the samples. Every vertical color strip represents a sample and every horizontal line represents the normalized expression values of one gene.

Reverse transcription and real-time PCR (qPCR)

The expression of *AKR1B1* was significantly ($P < 0.001$) lower in the Se-yeast group compared to the control (table 2.5). The results from the microarray experiment could be confirmed by qPCR in the 12 control and 14 Se-yeast samples. Slight differences between the two experiments within the groups occurred and the fold change with the qPCR measurement (-3.1) was comparable to the microarray experiment (-3.5). The same effect was observed for the genes *FADS1, GSTA1, PDZK1, SEPSECS*, and *TPMT*, which had significantly reduced gene expression. The increased expression of *GPX1, MRPS18A, PDZK1IP1*, and *SEPP* upon Se supplementation could also be verified by qPCR. A higher expression was detected compared to the microarray experiment, even though the effect was only pronounced with *GPX1*. The gene expression of *APOA5, GAL13*, and *SCLY* was not significantly altered upon Se-yeast supplementation with the qPCR measurement. The rate of genes from the microarray experiment, which could not be verified by qPCR was 0.23 and the correlation between the fold change in the microarray and the qPCR experiment was 0.95 of the confirmed genes.

Table 2.5: Results from microarray and qPCR measurement

Gene	Fc miroarray	Fc qPCR	T-test
GPX1	2.924	4.157	3.1E-06
AKR1B1	-3.526	-3.060	0.01
APOA5	1.855	1.217	NS
FADS1	- 1.332	- 1.224	0.036
GAL13	- 2.300	- 1.149	NS
GSTA1	- 2.170	- 1.779	2E-04
MRPS18A	-1.336	-1.130	0.047
PDZK1	- 1.966	-1.661	0.001
PDZK1IP1	1.414	1.426	0.016
P. / P.IP1	-2.613	-1.710	0.001
SCLY	- 1.256	- 1.067	NS
SEPP	2.045	2.092	1.5E-08
SEPSECS	- 1.330	- 1.106	0.038
TPMT	- 1.329	-1.167	0.040

Fc: Fold change in Se-yeast group compared to control, NS: not significant, P. / P.IP1: Ratio of single sample comparison PDZK1 / PDZK1IP1.
Fold change microarray: Normalization according to microarray experiment; control: 4 samples, Se-yest: 5 samples, qPCR: Control: 12 samples, Se-yeast: 14 samples.

Fatty acid analysis in the liver

Saturated fatty acids (SFAs) were significantly reduced by 4.7% in the Se-yeast group, which was due primarily to the significant reduction in C16:0 (-16.1%) (table 2.6). The concentration of C14:0 was low and there was no significant decrease of C18:0 in the Se-yeast group; therefore, these FAs did not influence the reduction of SFAs. Other SFAs, such as C15:0, C16:0iso, C17:0, C20:0, C22:0, C23:0, and C24:0, were increased in the Se-yeast group, but had only low concentrations or there was no significant difference between the groups (full list: Appendix A, table 7.3: Fatty Acid Analysis in the Liver). Monounsaturated FAs were not changed significantly, with the exception of eicosenoic acid (C22:1n-9) at a low concentration. PUFAs were significantly increased in the Se-yeast group by 13.8%, particularly due to the significant or by trend increase of linoleic acid (LA, C18:2n-6), α-linolenic acid (ALA, C18:3n-3), dihomo-γ-linolenic acid (DγLA, C20:3n-6), arachidonic acid (C20:4n-6, +23.4%), eicosapentaenoic acid (EPA, C20:5n-3), and tetracosapentaenoic acid (TPA, C22:5n-3). The n-3 and n-6 FAs were both significantly increased by 14.3% and 14.9%, respectively, but the ratio of n-6/n-3 was not altered. The Δ-5 desaturase ratios (C20:4n-6 / C20:3n-6 and 20:5n-3 / 20:4n-3) were significantly increased in the Se-yeast group by 21.7% and 14.9%, respectively. The *elongation of long chain fatty acid family member 5 (ELOVL5)* activity ratio (C18:4n-3 + C20:5n-3 + C18:3n-6 + C20:4n-6) / (C20:4n-3 + C22:5n-3 + C20:3n-6 + C22:4n-6) was only numerically decreased. In addition, the four single reaction activity ratios of *ELOVL5* were not significantly changed.

Table 2.6: Results from fatty acid analysis in the liver [g / 100g FAME]

Fatty acid	Control	St. dev.	Se-yeast	St. dev.	P-value
C16:0	20.11	1.92	16.88	2.73	0.002
C17:0	0.18	0.04	0.21	0.03	0.049
C18:0	22.43	1.86	22.42	2.05	NS
C18:2 n-6 c + t	15.20	1.37	17.13	1.92	0.007
C18:2 n-4[1,2]	0.04	0.01	0.04	0.01	0.024
C18:3 n-6	0.09	0.02	0.10	0.02	NS
C18:3 n-3	5.14	0.47	5.95	1.74	NS
C18:4 n-3[2]	0.06	0.01	0.08	0.08	NS
C18:4 n-6[2]	0.02	0.01	0.02	0.01	0.032
C20:0	0.10	0.01	0.10	0.01	NS
C20:2 n-6	0.53	0.08	0.59	0.06	0.042
C20:3 n-6	1.05	0.13	1.13	0.20	NS
C20:4 n-6	3.68	0.62	4.54	1.12	0.021
C20:3 n-3	0.40	0.09	0.48	0.09	0.025
C20:4 n-3[2]	0.29	0.02	0.27	0.05	NS
C20:5 n-3	5.21	0.77	5.76	1.49	NS
C22:0	0.06	0.02	0.07	0.02	NS
C22:1 n-9	0.02	0.01	0.03	0.01	0.047
C23:0	0.21	0.04	0.24	0.04	0.032
C22:4 n-6[3]	0.02	0.01	0.03	0.01	NS
C22:4 n-3[3]	0.05	0.01	0.05	0.01	0.020
C22:5 n-3[2,3]	2.43	0.51	2.76	0.60	NS
C24:0	3.76	0.82	4.73	1.49	0.046
C22:6 n-3	0.04	0.01	0.04	0.02	NS
∑ SFA	47.36	1.76	45.15	2.74	0.020
∑ MUFA	18.13	4.12	15.58	4.97	NS
∑ PUFA	34.42	3.20	39.17	4.85	0.006
∑ PUFA n-3	13.46	1.27	15.39	2.27	0.023
∑ PUFA n-6	20.50	2.25	23.56	2.97	0.007
n-6/n-3	1.48	0.14	1.50	0.16	NS
Δ5 desaturase n-6	3.48	0.31	4.00	0.71	0.026
Δ5 desaturase n-3	17.80	1.20	21.66	6.05	0.029
C20:4n-6 / C20:5n-3	0.71	0.12	0.79	0.12	NS (0.068)
ELOVL5	0.41	0.04	0.40	0.05	NS
C18:0 / C16:0	1.11	0.19	1.33	0.29	0.017
C18:2n-6 / C16:0	0.75	0.15	1.02	0.27	0.001
C18:3n-3 / C16:0	0.25	0.03	0.35	0.14	0.012
C20-24	18.35	2.53	21.36	4.62	0.046

St. dev.: Standard deviation, SFA: Saturated fatty acid, MFA: Monounsaturated fatty acid, PUFA: Polyunsaturated fatty acid, NS: not significant, $P > 0.05$, FAME: Fatty acid methyl esters. Number of samples: Control 14, Se-yeast 12, ELOVL5: (C20:4n-3 + C22:5n-3 + C20:3n-6 + C22:4n-6)/(C18:4n-3 + C20:5n-3 + C18:3n-6 + C20:4n-6).
[1]: Internal verification. [2]: According to Mondello [27] [3]: According to Hoffman [28]

Discussion

The broiler performance, BW, feed and water consumption, and feed conversion rate (FCR) were similar between the groups and comparable to those found in other studies [22]. The increased N retention in the Se-yeast group seems to be a result of reduced oxidative stress, which leads to reduced protein decomposition. The reduced protein decomposition will be discussed below in relation to the downregulated ubiquitin proteolytic and protein degradation pathways. An increase in digestibility of crude protein in male guinea pigs with a 0.1 ppm Se-supplemented diet has already been reported [29] and confirms our observation. The increase in AME could be due to better absorption and reduced oxidative stress in the intestine. The increased N retention might have an important influence on AME over the gross energy of the amino acids. The activity of GPX in the heparinized blood was significantly higher in the Se-yeast group than in the control group. According to Jaeschke [17], GPX activity is a reliable indicator of Se status. Due to a considerable Se concentration difference (0.551 mg/kg diet), the Se status of the broilers changed and could be confirmed. The downreguation of 77% of the significantly altered genes indicate a reduced RNA turnover (fig. 2.1). Even though the birds were from one single batch the Pearson correlation and the color pattern in fig.2.1 shows an elevated variance between the samples within the same group. This might be due to the use of commercial broilers, which were F2 hybrids with a high heterozygosity compared to inbred laboratory mice or rats, which have a high genetic homozygosity.

Selenoproteins and selenoamino acids

The five selenoproteins *glutathione peroxidase 1* (fold change (fc): 2.92, qPCR: 4.16, *GPX1*), *glutathione peroxidase 4* (fc: 1.43, *GPX4*), *selenoprotein P* (fc: 2.05, qPCR: 2.09, *SEPP*), *selenoprotein Pb* (fc: 1.30, *SEPPb*), and *apolipoprotein A-V* (fc: 1.86, qPCR: ns., *APOA5*) were significantly upregulated upon Se-yeast feeding. The GPX1, GPX4, SEPP, and SEPPb genes are related to the antioxidant system [30]. In a study by Fischer *et al.* (2001), a 13.9-fold downregulation of *GPX* was reported in rat livers due to Se deficiency [31]. The *GPX1* gene is ubiquitously expressed and detoxifies hydrogen and organic peroxides, including cholesterol and long-chain FA peroxides. Arthur [32] reported that the overexpression of *GPX1* in transgenic mice lines resulted in a reduction of ROS. The *GPX1* gene is specific to glutathione as a reducing substrate and the upregulation of *GPX1* in the liver was comparable with the two-fold activity increase in heparinized blood. The GPX4 enzyme detoxifies hydrogen and lipid peroxides with its monomeric structure and it may bind to a wider range of substrates than GPX1. Phospholipid hydroperoxides are removed by GPX4 but cannot be removed by GPX1 unless they are first metabolized by phospholipase A2, which was unchanged in our study. The protection of membranes against oxidative damage is principally maintained by GPX4 and this could have an important influence on the use of FAs. The expression of *GPX4* was not altered due to overexpression or knockout of *GPX1* [32]. In our study, the different fold changes support this observation. In the cell line RBL-2H3, which overexpressed *GPX4*, the conversion of arachidonic acid to intermediates in the lipoxygenase pathway was inhibited [33]. The FA analysis showed a significantly increased amount of arachidonic acid (table 2.6) in the liver of our broilers, but to what extent the reduced synthesis of leukotrienes contributed to the increased arachidonic acid concentration remains unclear.

It appears that SEPP is an antioxidant and Se transport protein. Knockout studies in mice tend to support the idea that SEPP is an Se transport protein that travels from the liver into the peripheral tissue [34]. The *APOA5* expression change was only significantly altered in the microarray analysis and the fold change was probably too low to be confirmed with qPCR. The APOA5 resides on high-density lipoprotein (HDL) and very low-density lipoprotein (VLDL) particles. Overexpression of *APOA5* in mice livers led to a reduction in plasma triglyceride to approximately one-third of the level in the control group; knockout mice, having no expression of *APOA5*, had a four-fold increase in plasma triglycerides. In addition, the VLDL particles increased in knockout mice and decreased in transgenic mice compared to the control. The overexpression of several other human apolipoprotein transgenes led to decreased triglyceride levels in mice [35]. The role of various apolipoproteins in triglyceride metabolism seems to be diverse. Epidemiological studies have shown that elevated triglyceride levels lead to a higher risk of coronary heart disease. Dichlberger *et al.* (2007) showed that the galline APOA5 binds to the major low density lipoprotein (LDL) receptor family of the laying hen. The LDL receptors (LDLR) mediate the endocytosis of LDL in the liver, which leads to a clearance of circulating LDL. The reduction in hepatic LDLR activity results in increased LDL cholesterol concentrations and the deposition of cholesterol in the body tissue [36].

The composition of Se-yeast seems to influence the expression of *Sep (O-phosphoserine) tRNA:Sec (selenocysteine) tRNA synthase* (fc: -1.33, qPCR: -1.18, *SEPSECS*), which is involved in selenocysteine-tRNA (SeCys-tRNA) synthesis [18, 37]. Only *SEPSECS* was downregulated in the synthesis pathway of SeCys-tRNA, which attaches phosphorylated selenide to the phosphorylated form Sep-tRNA, leading to SeCys-tRNA. *Selenocysteine lyase* (fc: -1.25, *SCLY*) was significantly downregulated in the Se-yeast group in the microarray experiment, but not in the qPCR experiment. It converts L-selenocysteine (SeCys) to selenide and L-alanine and functions as an Se delivery protein [38]. The Se in the feed additive Se-yeast is in the form of 83% seleno-L-methionine (SeMet), 5% SeCys, and lower percentages of other seleno amino acid analogs [18]. An explanation would be that the uptake of SeCys was higher than the requirement of the animals and led to the downregulation of *SEPSECS*, although the concentration of SeCys in Se-yeast was only 5%. According to Combs and Combs [39], a high concentration of free SeCys can be toxic and its accumulation is lethal. Bierla *et al.* (2008) reported SeMet and SeCys concentration ratios of 14 to 1.1 μg Se/g (12.7) in the breast muscles and 12 to 0.13 μg Se/g (92.3) in the leg muscles of broilers fed a diet supplemented with 6.9mg Se/kg feed (Se-yeast) [40]. These results show that an excess of SeCys might be degraded or converted to SeMet in leg muscles and that high amounts of organic Se are incorporated and stored in proteins as SeMet. The expression of *SEPSECS* and *SCLY* could be used as a biomarker for SeCys status in broilers. In the current experiment, optimal SeCys concentration seems to be reached because of the reduced SeCys-synthesis protein SEPSECS and the reduced SeCys-degradation protein SCLY.

The thiopurine S-methyltransferase (fc: - 1.33, qPCR: -1.67, *TPMT*) catalyzes the S adenosylmethylation of aromatic and heterocyclic sulfhydryl compounds like 6-mercaptopurine [41]. Ranjar *et al.* (2003) provide strong evidence that bacterial TPMT is involved in the methylation of inorganic and organic Se into dimethylselenide and dimethyldiselenide [42]. According to Terribas and Guijarro (2002), S-adenosyl-L-methionine serves as the S-methyl donor and is converted to S-adenosyl-L-homocysteine by TPMT [43]. The conversion of Se-Adenosylseleno-L-methionine to Se-Adenosylseleno-L-homocysteine has not yet been described in animals, but might be a possible reaction pathway of TPMT. However, the link between seleno-L-methionine and SeCys cannot be proven with certainty, TPMT seems to be a candidate gene to regulate the ratio between SeMet and SeCys.

Mihara *et al.* (2000) showed that pyridoxal 5'-phosphate (PLP) serves as a cofactor of *SCLY* [38] and Yuan *et al.* (2006) reported that PLP is required to carry out the Sep to SeCys conversion by *SEPSECS* [37]. Pyridoxal kinase (fc: -1.29, *PDXK*), which converts pyridoxal (Vitamin B6) to PLP [44], was downregulated in our study.

Glutathione metabolism

The *glutamate-cysteine ligase modifier subunit* (fc: -1.38, *GCLM*) and *glutathione S-transferase* (fc: -2.2, qPCR: -1.78, *GSTA1*) were downregulated; both proteins belong to the glutathione biosynthesis pathway. The GCLM is a subunit of the glutamate-cysteine ligase (GCL), which is the rate-limiting enzyme, and cysteine is the limiting substrate in glutathione synthesis. Glutathione homeostasis is regulated by a GCL feedback mechanism and by the glutathione redox system [45]. Mahmoud and Eden (2003) showed that the ratio of reduced glutathione to oxidized glutathione was significantly decreased in the livers of heat-stressed broilers [46]. In a different rat experiment, the GSTA1 activity and expression were decreased significantly in the liver from the control (0 mg Se/kg diet) to the group with a supplementation of 0.1 mg/kg sodium selenite in the diet [47]. These results indicate that gene expression might correlate with GSTA1 activity and is dependant on the level of Se supplementation. The *GSTA1* expression decrease was also confirmed by qPCR and might be a biomarker for the status of the antioxidant glutathione.

Downregulation of the ubiquitin proteolytic cycle

Twenty-four ubiquitin cycle - or related proteins were downregulated upon Se-yeast supplementation. The ubiquitin proteolytic pathway recognizes proteins that are oxidatively modified or damaged from oxidation by H_2O_2 or other ROS [48]. The accumulation of damaged proteins is cytotoxic and has been related to stress that occurs with aging or disease. It has been suggested that the ubiquitin proteolytic pathway specifically recognizes and removes oxidatively modified proteins [49]. The ubiquitin-activating enzyme E1 first activates ubiquitin, which is then transferred to the ubiquitin-conjugating enzyme E2 (fc: -1.28, *UBE2F*). The E2 enzyme can only transfer activated ubiquitin. Next, the ubiquitin is transferred to the protein substrate in the presence of E3 ubiquitin ligases, which are also required for the conjugation reaction as they play an important role in substrate recognition (fc: -1.65, *RNF11* / fc: -1.51, *MYLIP* / fc: -1.33, *FBXO9* / fc: -1.39, *FBXO22*) [50]. Myosin regulatory light chain interacting protein (MYLIP) is an E3 ubiquitin ligase that mediates the ubiquitination and subsequent proteasomal degradation of the myosin regulatory light chain [51]. The F box proteins (FBXO) are critical for the controlled and specific degradation of cellular regulatory proteins [52]. The topoisomerase I binding (fc: -1.29, *Topors*)

also seems to function like an E3 ubiquitin ligase together with specific E2 enzymes [53]. The ubiquitin proteases recognize the ubiquitin-tagged proteins and degrade them, like the 26S proteasome, which is a large multisubunit complex. It consists of a core proteinase, the 20S proteasome, and a pair of 19S regulatory particles. The following proteins are subunits of the 26S proteasome: fc: -1.28, *PSMC1* / fc: -1.29, *PSMA3* / fc: -1.20, *PSMD7*. Proteasome activator subunit 3 (fc: -1.33, *PSME3*) activates the proteasome 28S. The PSME3 protein amounts in UV light-irritated cells were elevated [54]. Ubiquitin carboxyl-terminal hydrolase 24 (fc: -1.44, *USP24*) is an ubiquitin-specific peptidase. Proteasome maturation protein (fc: -1.43, *POMP*) acts as a chaperone and helps to assemble the two half proteasomes (13-16S precursor) into the functional 20S proteasome [55]. Calcyclin binding protein (fc: -1.20, *CACYBP*) is a central component of the E3 ubiquitin ligase complex and is involved in β-catenin degradation [56]. Another protein related to the ubiquitin proteolytic pathway is the ubiquitin-fold modifier 1 (fc: -1.53, *UBFM1*). The entire cascade was downregulated from the ubiquitin-conjugating enzyme *E2* to the proteases.
The histone 2A family member V (fc: -1.34, *H2AFV*) is a mostly ubiquinated protein; 10–15% of this histone is ubiquinated within the cell [57]. The downregulation of the ubiquitin cycle might reduce H2A degradation and lead to a reduction of H2A synthesis in a feedback-regulated system. The downregulation of several ubiquitin cycle genes might also be related to the finding of Yoon *et al.* (2007) - that the feed efficiency of broilers improved linearly during the first three weeks, but had a quadratic improvement at week six when Se supplementation was introduced [58].
The reduced protein degradation might also have influenced the downregulation of six translation-related genes, *NACA, MRPL35, EIF1AX, MTIF2,* and the *mitochondrial ribosomal protein S18A* (fc: -1.34, qPCR: -1.13) [59]. The protein turnover seems to be reduced from the biosynthesis side as well as from the degradation side. These findings match our observation that N retention increased significantly in the Se-yeast group in the first sampling period. In the second sampling period, a slight increase could still be observed, but the significance level of $P < 0.05$ was not attained. The lower degradation rate of intracellular protein might lead directly to a higher apparent digestibility and a higher N efficiency, although FCR was not altered.

ROS a second messenger

Reactive oxygen species are free radicals that contain one or more unpaired electrons in their highest occupied atomic or molecular orbital. Hydrogen peroxide cannot directly be considered as a ROS. However, in the presence of transition metals, H_2O_2 can easily generate highly reactive hydroxyl radicals or other ROS. In low doses, ROS are also chemical mediators of the signal transduction processes involved in cell growth and differentiation [60]. The major sites of ROS generation are electron transport chains with leaking electrons that reduce O_2. Superoxide anions occur primarily towards the cytosolic side of the inner mitochondrial membrane. Three *NADH dehydrogenases* from complex I were downregulated (*NDUFB8,* fc: -1.45, *NDUFS6,* fc: -1.25 and *NDUFAF1,* fc: -1.25). Superoxide dismutase (u.n.) keeps the level of superoxide anions constant and only H_2O_2 can permeate the mitochondrial membrane [61]. Several genes that are related to intracellular ROS turnover and the signaling cascade were significantly changed. Silver *et al.*

(2003) reported that the hepatic overexpression of *PDZK1 interacting protein 1* in mice *(PDZK1IP1, MAP17)* resulted in a reduced amount of PDZ domain containing 1 (PDZK1) in the liver, probably due to post-transcriptional changes [62]. In our experiment, we observed those alterations on the level of mRNA; *PDZK1IP1* expression (fc: 1.41, qPCR: 1.43) was significantly increased and *PDZK1* expression (fc: -1.97, qPCR: -1.66) was significantly decreased. In the microarray and qPCR experiments, alterations reached significance and the expression relation *PDZK1 / PDZK1IP1* was also significantly altered. Further, Silver *et al.* (2003) reported a simultaneous two-fold increase of blood HDL in *PDZK1IP1* overexpressing mice. They also mentioned that mice with targeted disruption of the *PDZK1* gene have increased plasma HDL levels [62]. The link to the upregulated gene *APOA5* and the altered FA composition in our study remains unclear and the measurement of HDL in the blood of broiler chicks might be beneficial in further studies. In a second study with A375 cells, the overexpression of *PDZK1IP1* led to increased tumorgenic properties, like anchorage-independent growth and increased migration. The same authors also mentioned that a large variety of human carcinomas showed an increased *PDZK1IP1* mRNA level. The *PDZK1IP1*-overexpressing clones showed a 30% increase in ROS levels. It was assumed that *PDZK1IP1* generated ROS functions as an intracellular signal, inducing a growth-related genetic program, which results in tumorgenic properties. On the other hand, the colony formation efficiency was reduced in the presence of the antioxidant glutathione [63]. It remains unclear how the ROS reduction, due to the supplementation with Se-yeast, influences PDZK1IP1 expression. Yet PDZK1IP1 seems to be an essential gene in ROS signaling and tumorgenesis.

It has been shown that H_2O_2 stimulation leads to tyrosine phosphorylation of fms-related tyrosine kinase 4 (fc: + 1.26, *VEGFR3 / FLT4*) in HMEC-1 cells. VEGFR3 signaling induces the activation of the PI3-kinase/Akt pathway [64]. The PI3-kinase/Akt pathway plays an essential role in cell survival when challenged by oxidative stress [65]. Further, Akt phosphorylates caspase-9 (fc: + 1.30, *CASP9*) and thereby inhibits apoptosis [66]. The upregulation of the genes *FLT4,* and *CASP9* might increase sensitivity to ROS. Several other apoptosis and programmed cell death genes *(PDCD6, AIFM1, PDCD5, BRWD2, PDCD11* and *ATN1)* were significantly downregulated. In general, these findings would indicate a lower apoptosis rate in liver tissue when an animal's diet is supplemented with Se-yeast. Further, Serine/threonine kinase 25 (fc: -1.63, *STK25*) is an oxidative stress-related protein activated by autophosphorylation and seems to be a biomarker for oxidative stress [67].

The activity of aldo-keto reductase family 1, member B1 (fc: -3.53, qPCR: -3.06, *AKR1B1*) was significantly reduced by 18.5% due to Se-yeast supplementation. The substrate glycerladehyde in the activity assay can be reduced by enzymes such as AKR1A1 and AKR7A2, which have different catalytic activities [68]. Additionally, the expression of AKR1A1 was not significantly altered upon Se-yeast supplementation and the expression of AKR1B1 (AKR1B1/A1: 0.86) was lower compared to AKR1A1. The difference in activity between the groups was low because the catalytic activity of AKR1B1 seems to be covered by the background reactions from other enzymes. In their review, Jin and Penning [69] describe AKR1B1 as an enzyme, which reduces aldehydes and ketones to primary and secondary alcohols. In addition, reactive aldehydes and/or dicarbonyls are detoxified by

AKR1B1. ROS also target PUFAs (e.g., linoleic acid and arachidonic acid) that form lipid hydroperoxides, which finally decompose to form reactive lipid aldehydes (4-hydroxy-2-nonenal (4-HNE) and 4-oxo-2-nonenal (4-ONE)). The lipid aldehydes 4-HNE and 4-ONE can react with bases in DNA to form etheno- and heptano-etheno-DNA adducts, which can be highly mutagenic. AKR1B1 can detoxify 4-HNE to 1,4-dihydroxy-2-nonene and 4-HNE can be converted enzymatically or nonenzymatically with glutathione to 3-glutathione-4-hydroxynonene (3-GSH-4-HNE) and then to 3-glutathiuonyl-1,4-dihydroxynonanol. When sufficient GSH is present, the second pathway is preferred and AKR1B1 seems to be the only enzyme that can detoxify 3-GSH-4-HNE. Other toxic metabolites like acrolein and 1,2-naphthoquinone are disposed of by AKR1B1. *AKR1B1* is induced by osmotic stress and other *AKR1* genes are also regulated by the antioxidant response element [69]. *AKR1B1* expression might be a biomarker for reactive lipid aldehydes, which are generated by ROS challenge.

Lipid metabolism

Selenium modulates the composition of FAs in the liver and led to the reduction of the expression of the *fatty acid desaturase 1* (fc: -1.33, qPCR: -1.22, *FADS1*), the *elongation of long chain fatty acid family member 5* (fc: -1.23, *ELOVL5*), and the *acyl-CoA synthase long-chain family member 5* (fc: -1.75, *ACSL5*). ACSL5 performs the ligation of palmitic acid with CoA to the starting substrate palmitoyl-CoA (C:16:0-CoA) for PUFA biosynthesis. The level of C16:0 was significantly reduced and the ratio of C18:0 to C16:0 respectively the ACSL5 conversion rate was significantly increased in the Se-yeast group. The C18:0 concentration was not altered in the Se-yeast group; therefore, the reduced expression of *ACSL5* could not be directly attributed to the FA profile. It has to be considered that the reduced C16:0 concentration could be the result of a lower synthesis rate or that ACSL5 was not the rate-limiting factor in the conversion of C16:0 to C18:0. However, it must be mentioned that PUFA concentration was increased in the Se-yeast group, which would explain a shift from C16:0 to longer and unsaturated FAs. Reinartz *et al.* (2010) reported an upregulation of ACSL5 when C18:1n-9, C18:2n-6, and C16:0 were added to HepG2 cells [70]. In our *in vivo* experiment, the concentration of C16:0 was decreased, C18:1n-9 was unchanged, and C18:2n-6 was increased. It seems that the 16% decrease of C16:0 led to the downregulation of ACSL5 and could not be recovered by the 12% increase of C18:2n-6.

ELOVL5 is responsible for the first steps in the elongation of C18:4n-3 to C20:4n-3, C20:5n-3 to C22:5n-3, C18:3n-6 to C20:3n-6, and C20:4n-6 to C22:4n-6 [71]. The FA analysis showed no significant change in the conversion rate of EVOVL5 or in the single reaction activity ratios of the n-3 and n-6 FAs. The elongation and desaturation of the FA chain requires several other enzymes and ELOVL5 does not seem to be the rate-limiting enzyme in this cascade. FADS1 desaturates the C5 bond *(Δ-5 desaturase)* and converts C20:4n-3 to C20:5n-3 and C20:3n-6 to C20:4n-6. Both Δ-5 desaturation rates of the n-3 and n-6 FAs were significantly increased. FADS1 is one of the rate-limiting enzymes in the biosynthesis of PUFAs [72]. The reduction of *FADS1* expression seems to be due to the significant increase of PUFA. Cho *et al.* (1999) reported a reduced expression of *FADS1* in the liver of rats when their diet contained high amounts of PUFA [72]. In our study,

PUFA concentration was significantly increased in the liver and this seems to be the reason for the decreased expression of *FADS1* in the Se-yeast group. The reduced use of PUFA due to reduced oxidative stress seems to influence the gene expression of *FADS1* as a secondary effect. In a study by Schafer *et al.* (2004), the high Se-treated rats (sodium selenite) had significantly higher FA concentrations of C16:0, total n-3 PUFA, C22:6n-3, and total C20-C22. C18:2n-6, total n-6 PUFA, and the ratio of n-6/n-3 were significantly lower in the Se-supplemented group [19]. In the present study, only total n-3 PUFA and C20–C24 were significantly increased. In contrast, C16:0 was significantly reduced, C22:6n-3 and the n-6/n-3 ratio were unchanged, and C18:2n-6 and the total n-6 were significantly increased. The differences in the results may be attributed to Se concentration, the inorganic Se source, or the FA content in the diet or the species. However, it shows the high variability of the FA concentration. The increase of *GPX* activity and the possible lower ROS content in the cells might reduce the absorption of PUFA, especially EPA, DHA, and arachidonic acid, which are vital components of phospholipids in membranes. The arachidonic acid concentration was significantly increased by 23% in the Se-yeast group. The concentrations of the different products and educts might influence the enzymes, as well as the entire cascade, depending on the concentrations of several initial FAs [73]. In the complex cascade of PUFA biosynthesis, weak and insignificant expression alterations of several genes might influence the FA content without being detected by microarray analysis. Schafer *et al.*(2004) reported an increase of triacylglycerols (TAGs) in the livers of Se-adequate fed rats and suggested that TAGs serve as expendable FA reserve for the synthesis of phospholipids [19]. If there was reduced use of FAs for phospholipids due to reduced oxidative stress and better protection of PUFA, this would explain the increased concentration of PUFA in the livers in our study.

Microsomal triglyceride transfer protein (fc: - 1.55, *MTTP*) enables the secretion of VLDLs by the liver and is regulated over the transcription rate by insulin. Raabe *et al.* (1999) showed that the knockout of *MTTP* in the liver of mice led to the absence of TAG in the endoplasmatic reticulum (ER), where VLDLs are assembled for later secretion [74]. It seems that the export of TAGs over VLDLs is reduced due to the downregulation of MTTP and the supplementation of Se-yeast. This would be in agreement with the increased expression of APOA5 and a possible decrease of TAGs in the plasma discussed previously. *In vitro* experiments showed that the inhibition of *MTTP* is mediated over the MEK1/2 ERK cascade [75].

Vesicular transport

Myosin IB (fc: -1.62, *MYO1B*) controls protein cargo traffic in multivesicular endosomes together with actin filaments [76]. Other proteins related to vesicular transport are syntaxin 7 (fc: -1.36, *STX7*) and syntaxin 2 (fc: - 1.22, *STX2*). They are related to intracellular vacuolation and are implicated in the targeting and fusion of intracellular transport vesicles. Sorting nexin 4 (fc: - 1.71, *SNX4*) is involved in intracellular trafficking; vacuolar protein sorting 24, 26B, 36, and 53 homolog (fc: - 1.29, *VPS24* / fc: - 4.01, *VPS26B* / fc: -1.30, P, *VPS36* / fc: - 1.41, *VPS53*) are related to vacuolar transport, and BET1 homolog (fc: -1.28, *BET1*) encodes a golgi-associated membrane protein that participates in vesicular transport from the endoplasmic reticulum (ER) to the Golgi complex. Other downregulated genes influencing vacuolar and ER transport are the genes from and related to the RAB family (fc: -1.28, *RAB1A* / fc: -1.46, *RAB33B* / fc: -1.24 *GDI2*, fc: -1.45, *RIN2*), the Sec family (fc: -1.28, *SEC23A* / fc: -1.26, *SEC61G* / fc: -1.33, *SEC63*), and others (fc: -1.43, LMAN1 / fc: -1.57, RHBDF1) [77-79]. SEC61G is part of the Sec complex, together with two other subunits. Sec is a membrane component of the ER, which forms a pore in the membrane through which the nascent polypeptide is translocated. SEC63 can form a complex with Sec and seems to be related to the backward transport of ER proteins that are subject to ubiquitin-proteasome-dependent degradation [80]. The two genes *SEC61G* and *SEC63* could be related directly to the reduced proteasome degradation pathway and the increased N retention of the broiler chicks. All the downregulated genes related to vesicular transport and to the secretory pathway might also have contributed to the altered FA profile in the livers and to increased AME.

Nell *et al.* (2007) generated a gene expression profile from the livers of mice, fed low and an adequate dose of vitamin E. Several vesicular transport and related transcripts were altered. They concluded that the antioxidant effect of vitamin E led to the increased protection of unsaturated FAs. Vesicular transport is dependent on intact membranes, and with lower oxidative stress, the damage and therefore the renewal and repair of membranes is reduced [81]. This might affect the feedback mechanism and the expression of vesicular transport and related proteins. Another explanation might be that lower ubiquitination and degradation of proteins results in decreased protein biosynthesis, and subsequently, decreased protein transport within cells.

Protein maturation

Newly synthesized and misfolded proteins can be folded or refolded with the help of chaperones. The refolding is required due to damages in the protein structures. *Chaperonin containing TCP1 (TRiC) subunits* (fc: -1.281, *CCT4* / fc: -1.272, *CCT5*) were downregulated. TRiC are molecular chaperones that bind directly to proteins to prevent aggregation and promote protein folding. Approximately 9–15% of newly synthesized proteins are estimated to be associated with TRiC, among them are *actin* (fc: +1.436, *ACTG1, gamma* / fc: -1.332, *ACTL6A*), *tubulin* (fc: -1.444, *TUBB6*), and Myosin (fc: -1.616, *MYO1B*) [82].

Calnexin (fc: -1.337, *CANX*) is a transmembrane protein and contains an ER localization signals. Newly synthesized glycoproteins bind transiently and selectively to calnexin in the ER. CANX is a molecular chaperone that prevents aggregation and export of the incompletely folded proteins from the ER and the subsequent degradation by the ubiquitin cycle [83, 84].

N-glycanase 1 (fc: -1.322, *NGLY1*) interacts with the ubiquitin complex and deglycosylates misfolded glycoproteins from the ER. The exact location of *NGLY1* has not yet been determined; it may be on the surface of the ER or in the cytosol. The deglycosylated proteins are then targeted for degradation by the proteasome and the ubiquitin pathway [85]. The downregulation of several maturation related proteins might be closely related to the downregulation to the ubiquitin proteolytic pathway and the reduced amount of oxidized proteins. The decrease might also be influenced by the reduced vesicular transport, which is important for the maturation of newly synthesized protein.

Conclusions

The main effect of Se supplementation was reduced oxidative stress due to the increased activity and expression of GPX enzymes. This seems to be the reason for the downregulation of the ubiquitin pathway and the subsequent increase in N retention. The N retention increase is an indication that the protein turnover rate could be reduced because of the decreased amount of damaged protein. On the side of protein biosynthesis, the N retention seems to be influenced by genes related to the translation and transcription pathway. The altered FA profile and the increase in PUFA indicate reduced oxidation of FAs, leading to the reduced repair of lipid bilayers. However, it remains unclear to what extend reduced protein or FA use influenced the reduced expression of genes related to vesicular transport. It seems that both effects had an impact on the decreased necessity of vesicular transport. Changes in the biosynthesis of selenoamino acids indicate the unequal use of SeMet and SeCys. Additional research on the composition of organic Se sources could also be beneficial in human nutrition.

According to Rayman [10, 18], Se intake in human nutrition seems to be suboptimal with respect to disease risks. The supplementation with organic Se in the feed of broilers and other farm animals is a possible strategy to increase human Se intake.

In our study, we had the impression that the use of four or five microarrays per group is probably the lower limit for nutrigenomics. But the use of microarrays seems to be a very efficient starting point for the study of pathway regulation, protein interaction and the alteration of RNA levels and proteins in live animals. Further, the study of nutritional trace elements seems to be an interesting approach to fundamental research in living animals by influencing gene expression without harming the animals. Therefore, using live animals, we could confirm the results of several authors, which were based mainly on cell cultures and *in vitro* techniques. It must be mentioned that the impact of altering the trace element content in animals' diets is minimal and due to the screening of all expressed genes, coverage on all known pathways is possible.

3. *Fusarium* mycotoxin-contaminated wheat containing deoxynivalenol alters the gene expression in the liver and the jejunum of broilers

Based on:

Bruno Dietrich, Stefan Neuenschwander, Benjamin Bucher and Caspar Wenk

Animal Journal (published) [86]

Abstract

The effects of mycotoxins in the production of animal feed was investigated using broiler chickens. For the feeding trial, naturally Fusarium mycotoxin-contaminated wheat was used, which mainly contained deoxynivalenol (DON). The main effects of DON are a reduction of the feed intake and reduced weight gain of broilers. At the molecular level DON binds to the 60S ribosomal subunit and subsequently inhibits protein synthesis at the translational level. However, little is known about other effects of DON e.g. at the transcriptional level. Therefore a microarray analysis was performed, which allows the investigation of thousands of transcripts in one experiment. In the experiment, 20 broilers were separated into four groups of five broilers at day 1 after hatching. The diets consisted of a control diet and three diets with calculated, moderate concentrations of 1.0, 2.5 and 5.0 mg DON/kg feed, which was attained by exchanging uncontaminated wheat with naturally mycotoxin-contaminated wheat up to the intended DON concentration. The broilers were held at standard conditions for 23 days. Three microarrays were used per group to determine the significant alterations of the gene expression in the liver ($P<0.05$), and real-time PCR was performed in the liver and the jejunum to verify the results. No significant difference in body weight, feed intake or feed conversion rate was observed. The nutrient uptake into the hepatic and jejunal cells seemed to be influenced by genes; SLC2A5 (fc:-1.54, DON2.5), which facilitates glucose and fructose transport and SLC7A10 (fc:+1.49, DON5), a transporter of D-serine and other neutral amino acids. In the jejunum the palmitate transport might be altered by SLC27A4 (fc:-1.87, DON5) and monocarboxylates uptake by SLC16A1 (fc:-1.47, DON5). The alterations of the SLC gene expression may explain the reduced weight gain of broilers chronically exposed to DON-contaminated wheat. The decreased expressions of EIF2AK3 (fc:-1.29, DON2.5/5) and DNAJC3 (fc:-1.44, DON2.5) seem to be related to the translation inhibition. The binding of deoxynivalenol to the 60S ribosomal subunit and the subsequent translation inhibition might be counterbalanced by the down-regulation of EIF2AK3 and DNAJC3. The genes PARP1, MPG, EME1, XPAC, RIF1 and CHAF1B are mainly related to single-strand DNA modifications and showed an increased expression in the group with 5 mg DON/kg feed. The results indicate that significantly altered gene expression was already occurring at 2.5 mg DON/kg feed.

Implications

The mycotoxin deoxynivalenol is a frequently detected compound produced by molds, which grow on wheat and other cereals during pre-harvest. The main effects of DON in broilers are a reduced feed consumption and weight gain. The measurement of the gene expression in the liver gave indications on the function mechanism of DON. Genes related to altered nutrient uptake to the cell, detoxification, altered protein synthesis rate and DNA repair mechanisms could be detected. With the description of new pathways a better understanding of the effects of DON on broiler chicks' metabolism can be expected.

Introduction

About 25% of the world's food crops are contaminated with mycotoxins. The most frequently occurring and widespread mycotoxin is deoxynivalenol (DON) in feedstuffs for poultry in European countries [87]. The mycotoxin-contaminated wheat used in this study contained mainly DON in addition to a very low contamination with T-2 and zearalenone in relation to maximal guidance values in the European Union. DON is a type B trichothecene, a secondary metabolite produced by Fusarium species mainly from F. graminearum and F. culmorum. Due to the growth of the fungi on maize, wheat and other cereals, poultry feed can contain DON [11, 87]. In the European Union the maximal permitted value for feed contamination is 5 mg DON/kg feed in poultry nutrition. At that level, no damage to the birds is expected [88]. However, decreased feed intake, weight gain and reduced peripheral blood monocytes have been observed in broilers fed 9.3 mg/kg DON in the grower phase [89]. DON fed to pigs either in a purified form or in the form of naturally contaminated corn led to reduced weight gain and feed intake. Until the end of the experiment, the animals receiving the purified DON recovered from the growth depression in contrast to the animals receiving naturally contaminated corn. The differences between naturally DON-contaminated feed sources and the feed spiked with purified DON might be caused by fungal components present in the naturally mycotoxin-contaminated feed [90]. We decided to use naturally mycotoxin-contaminated wheat for the experiment to reproduce the effects that occur in the commercial broiler production. It was suggested that the DON tolerance of broilers results from poorer toxin bioavailability and rapid elimination from the body. DON is rapidly absorbed in the upper part of the gastrointestinal tract and is also quickly cleared from the chickens' body. It has been suggested that an efficient clearance takes place due to a hepatic or renal first-pass effect in chickens. The oral administration of 14C-DON to chickens showed high radioactivity in the liver and bile, with more than 90% of the original label occurring in the excreta before 48-h [12]. Another experiment showed that in hens administered 3H-DON, an overwhelming amount of radioactivity was excreted in the urine, demonstrating the absorption of DON by the digestive system [91]. The main DON metabolite detected in urine and feces was de-epoxy DON (DOM-1) [92]. Therefore the liver, as a direct target of DON, was chosen to analyse the gene expression with microarrays. It has been suggested that cells and tissues with high protein turnover rate, such as the liver and small intestine, are most affected by DON [12]. It was also observed that fast-growing broilers are more susceptible to DON than laying hens [11]. Pestka [93] claimed that DON is metabolised in all species, and that it does not accumulate. The main effect of DON is the binding to the 60S ribosomal subunit and the subsequent prevention of polypeptide chain initiation or elongation [94]. Further, DON affects the immune system either as an immunosuppressive or immunostimulative, depending on dose and exposure regime. The immunosuppression was explained mainly by the binding capacity of DON to the ribosomes and the inhibition of protein synthesis [95]. However, there is little or no information available about the involvement of DON in pathways of immune suppression, nutritional uptake or detoxification within the living animal. Most of the studies at the cellular level focused on the acute effects of DON administration. Long-term effects of DON-contaminated cereals seem to be important for animal production due to a regular occurrence of DON-contaminated feed.

The microarrays used in our experiment contained 37’000 probe sets corresponding to more than 28’000 genes. Especially when little is known about the molecular effects of a substance, the use of microarrays as a screening method provides an invaluable starting point. The sequence of the probe sets was designed according to the sequence information from all transcribed genes, including the ones with an unknown function. The profiling of gene expression with microarrays should help to gain insight into the effects and metabolism-related genes upon DON administration. Additionally the expression of genes, with a biological relevance to DON administration, was verified with reverse transcription real-time PCR in the liver as well as in the jejunum, to cover the main target organs of DON. Possible marker genes and affected pathways should be identified to achieve a better understanding of how chronic and low doses of mycotoxin-contaminated feed uptake affects liver and the small intestine cells in broilers. In addition, the concentration at which DON influences the gene expression in the liver should be reviewed.

Material and methods

Experimental design, birds and diets

Twenty one-day-old male ROSS 308 broiler chicks were obtained from the commercial hatchery Erb Brüterei AG (Oberdiessbach, Switzerland). The birds were weighed at the beginning of the experiment and separated into four treatment groups according to similar group weight. Five broilers per treatment group and cage were raised together for 23 days. A standard diet (Table 3.1) was formulated according to the National Research Council [96] and prepared for all groups in the same process. DON was added to the basic feed mixture of the treatment groups in the form of mycotoxin-contaminated wheat (DON: 52 mg/kg). Deoxynivalenol was determined in the diets with LC-MS/MS; the other mycotoxins were determined in the contaminated wheat due to their low concentration, which was close to or below the detection level in the diet. T-2 toxin and zearalenone concentrations of the contaminated wheat were determined with LC-MS/MS, and ochratoxin A; fumonisin B1 and B2 and aflatoxin B1, B2, G1 and G2 were determined with HPLC-FLD in an external laboratory (UFAG, Sursee, Switzerland). The pelleted feed and water were provided *ad libitum* from day 1 onward. The animal experiment was performed according to the regulations of the cantonal veterinary office in Zurich.

Table 3.1: Broiler diets

	Control	DON1	DON2.5	DON5
DON [mg/kg]	0	1	2.5	5
Ingredients %				
Corn		17.5		
Soybean meal 48%		25.0		
Potato protein		2.5		
Rapeseed oil		5.0		
Salt		0.15		
DL-methionine		0.26		
Limestone grit		0.90		
Lysine-HCL		0.17		
DCP 38/40		1.20		
Na-bicarbonate		0.30		
Celite 545		1.52		
Premix1		0.50		
Cont. wheat	0	1.92	4.80	9.62
Uncont. wheat	45.00	43.08	40.19	35.39
Analyzed parameters				
DON mg/kg	0.32	0.88	2.21	4.42
T2-toxin [µg/kg]	0	1.0	2.5	5.1
ZON [µg/kg]	0	6.7	16.8	33.7
Dry matter [%]	90.9	90.9	91.2	91.2
G.E. [MJ/kg]	15.85	15.85	15.94	15.74
Crude fat [%]	6.81	6.97	7.03	6.97
Crude protein [%]	20.76	20.73	20.86	20.88

[1]Premix: Diet from Aeschbacher *et al.* (2005) [22].
Contaminated wheat (52 mg DON/kg), Uncontaminated wheat (0.58 mg DON/kg).
Maximal allowed contamination per kg feed of poultry in European Union (2006) (percentage in group DON5 of maximal allowed concentration): DON: 5 mg/kg (88.5%), T-2: 150 µg/kg (3.4%), zearalenone (ZON): no recommendations. Mycotoxins below the detection limits: Aflatoxin B1, B2, G1, G2 (1.3 µg/kg feed), ochratoxin A (1 µg/kg feed), fumonisin B1, B2 (10 µg/kg feed).

The total cage weight and feed intake were monitored weekly. Total water intake per cage was recorded for four days a week. Feed samples were taken from the principal components, the control and the treatment diet before and after pelleting and on days 1 and 23 of the experiment. All the feed samples were stored in a 4 °C cooler. Excreta samples were taken from day 8 to day 11 and from day 15 to day 18 from all four cages and stored at -20°C. Dry matter, ash, gross energy, crude fat and crude protein content were determined in all feed and excreta samples (Table 3.1). Apparent metabolisable energy intakes (AME) and N retention were calculated according to Scott and Hall [20]. After 23 days, broilers were weighed individually and slaughtered by cervical dislocation. Liver samples were removed immediately and washed, and from the middle part of the left liver lobe, a slice of about 1 cm width was cut and frozen in liquid nitrogen. In a similar way a slice of the jejunum was collected; first 7 cm below the angle of Treitz, a 4-cm-long piece from the jejunum was cut out and then the lumen was opened and rinsed with 0.9% NaCl before it was also frozen in liquid nitrogen.

High-density oligonucleotide array hybridization

Total RNA was isolated from 12 frozen liver samples with TRIZOL reagent (Invitrogen Basel, Switzerland). From all four groups, three samples were selected according to the bird's weight. The heaviest and lightest birds were excluded. The samples were homogenised in a mortar filled with liquid nitrogen. After isolation, RNA concentration and purity were determined with an ND-1000 (NanoDrop Technologies, Wilmington, DE, USA) and further cleaned up with the RNeasy Mini Kit (Qiagen, Hilden, Germany). Before the reverse transcription, the RNA integrity was determined using a Bioanalyzer 2100 (Agilent Technologies, Basel, Switzerland). The integrity number was calculated using the Bioanalyzer 2100 software with a value between 1 (for totally degraded RNA) and 10 (for undegraded RNA). For the microarray experiments, only samples with a value higher than 9 were taken. Labelled cRNA probes were generated according to the protocol from Affymetrix using the GeneChip® One-Cycle cDNA Synthesis Kit (Affymetrix, Santa Clara, CA, USA). The procedure consisted of a reverse transcription of 7μg total RNA with a T7 oligo (dT) promoter primer and a second-strand cDNA synthesis followed by an in vitro transcription to biotin-labelled cRNA. Finally the labelled cRNA was fragmented and hybridised at 45°C for 16 hours on the Chicken Genome Arrays (Affymetrix, Santa Clara, CA, USA). For every sample a single microarray was used. The arrays were washed and stained using the GeneChip Fluidics Station and scanned with a GeneChip Scanner 3000 according to the manufacturer's protocol (Affymetrix, Santa Clara, CA, USA). The GeneChip operating software (GCOS, Affymetrix, Santa Clara, CA, USA) controlled the scanner and the fluidics station. The data were analysed with Microarray Suite version 5.0 (MAS 5.0) using Affymetrix default analysis settings and global scaling as a normalisation method. The image data on each individual chip was scaled to target intensity 500.

Reverse transcription and real-time PCR

To verify the results from the microarray experiment, the total RNA was extracted from the remaining liver and the jejunum samples. RNA extraction, quantification and quality control were performed equally to the microarray samples. The experiments for 20 samples per tissue were conducted in two steps: first, the reverse transcription (RT) was performed using the High-Capacity cDNA Reverse Transcription Kit (Applied Biosystems, Rotkreuz, Switzerland), and then the real-time PCR (qPCR) was conducted with adjusted amounts of cDNA. The genes for the qPCR were selected after the gene expression analysis and the literature review, depending of their possible biological relevance in relation to mycotoxin-contaminated feed. Primers were designed with Primer Express® 3.0 software (Applied Biosystems, Rotkreuz, Switzerland) on two adjoining exons in order to exclude a possible amplification due to genomic DNA contamination. The two endogenous controls mitochondrial ribosomal protein S18A (MRPS18A) and ribosomal protein S17 (RPS17) turned out to perform the best for adjusting the cDNA concentration, compared to the traditional and other selected control genes such as: GAPDH, β-actin, DHRS7, RPS28, APOB and LDHA. Genes and the corresponding primer pairs are indicated in Table 3.2. The 20-μl qPCR reaction contained 500nM of forward and reverse primers, 100 ng of sample cDNA and 10 μl Fast SYBR Master Mix (2X) (Applied Biosystems, Rotkreuz, Switzerland). The PCR amplification was performed on the 7500 Fast real-time PCR System with SDS Software 1.3.1 (Applied Biosystems, Rotkreuz, Switzerland) and repeated four separate times. The cycling program was 20 sec at 95°C and then 40 cycles of the following: denaturation for 3 sec at 95°C, annealing and extension at 60°C for 30 sec. The results were evaluated with the comparative Ct method (2-ΔΔCt) with the SDS Software.

Table 3.2: Primer sequence for real-time PCR

Gene	Forward	Reverse
MRPS18A	GGGAAGCGGCTGGGTTT	ATTATGGTCGTGTTGCCTTCAGT
RPS17	TGGATGCGCTTCATCAAGTG	TCATCCCCCAGCAAGAAAGT
AKR1B1	TTAAAGAGATTGCAGCCAAGCA	TCACGTTTCTCTGGATGTGGAA
CASP1	GCAGGAGATGTTCCGAAAGG	ACTTCTTCAGCATTGTAGTCCTCTCTT
EIF2AK3	CCTGTTCTGCCTCATCGTCAT	ACTCAGGGAAGGTGAAGTACATCTG
EP300	GGCTGAGCTGTTGGCATAGC	TCAACCATGAATTCCCAAAACTC
EXOSC9	CGCGTTCCTCTCGTTCATTT	CCTATTTGTGTCAGTTTTGCCTTCT
IFT57	AGGATTAGTGGGCAGTGCAAA	GCTGAAGCTGCTGAGCTACGA
LAPTM4B	CAGGACTACCTTCGCCAGCT	TTCCTTGTAAGGAAAGTTGCTGG
MIA2	GGAATGTGAGACGGCAATGA	TCCCCTGTTTTGAAGCTCAGA
MAPKAPK3	TGGTCTCTGGGTGTCATCACA	AAATAGCTTGTCCAGTGTTCGAGTAG
TP53I3	TGGCCAAGCTGAATATGTTACAGTA	GCTGCAGCCTGAATAAAAGTCA
SLC2A5	ACTGCAGCAACAATGGAGAAGA	GGTGGGACCTCTGGTCAACA
SLC7A10	TGTAAATATATCCTGAATGCGTGTTG	CTAGGATTCTCTCCATGGTGTGTTT
SLC16A1	GCAGGGTTCAGGTTAAATGCA	CACCGTGGAGGAGCTCTACTTC
SLC27A4	TCCTGCCTTCCCCTCAATG	CGCATCCTCAACCTGACAGA
STK39	ACTCGTAACAAAGTAAGGAAAACATTTG	CCTCGCACCTGCTCCATTA
TJP1	TCTTGCTGTGGCTACAACAGTGT	AATGCTGTGCCTAAAGCCATTC
XRN1	ACTTCACTGAAGTGCTCACGATCA	GATATTTCTCAAAGCGCTTTAAATTCAG

F.: Forward primer. R.: Reverse primer. 5' - 3' direction.

Statistical analysis

The alterations in the expression profiles between the control and the treatment group were analysed using GeneSpring 7.3 software (Agilent Technologies, Basel, Switzerland). Normalisation was carried out with GeneSpring 7.3 and the default settings in the following way: values below 0.01 were set to 0.01. Each measurement was divided by the 50th percentile of all measurements in that sample. And each gene was divided by the 50th percentile of its measurements in all samples. The pre-filtering was performed according to Pepper et al. (2007) [97] the following way: The genes had to be present or marginal in all samples, and the raw value had to be above the 50th percentile, of all present and marginal genes, in all samples of a group to remove genes with a low signal strength. The gene lists from the four groups were merged together, and Student's t-test was performed between the control and each treatment group (P<0.05). Then, genes were grouped in seven subgroups: first, genes with significant changed expression in only one group (DON1, DON2.5, DON5) compared to the control group; then, genes within two groups (DON1/2.5, DON2.5/5, DON1/5) compared to the control group and genes within all three groups (DON1/2.5/5) compared to the control. Finally, for every subgroup a t-test was performed and the p-values indicated in Tables 8.1 and 8.2 (appendix B). The fold changes were listed for every comparison. The final gene list was separated into groups of up- and down-regulated genes. The microarray raw data have been deposited into the Gene Expression Omnibus (National Center for Biotechnology Information http://www.ncbi.nlm.nih.gov/geo) as accession number GSE25185. Genes with an average raw value above the 80th percentile, of all present and marginal genes, in the group with the significantly changed expression were indicated with an asterisk (Tables 8.1 and 8.2). The average raw value was indicated to estimate the importance of a gene in cellular processes. All the other measured parameters were analysed using the Student's t-test to determine significance (P<0.05).

Results

Diet characterization and performance of broilers

The DON concentration of the control diet was 0.32 mg/kg feed, which was close to the detection level of 0.2 mg DON/kg. The DON concentration in DON1 was 0.89 mg/kg, 2.21 mg/kg for DON2.5 and 4.43 mg/kg for DON5. The calculated T-2 toxin and zearalenone concentrations in group DON5 were low, with values of 5.1 µg/kg and 34.6 µg/kg feed, respectively. The other measured mycotoxin concentrations were below the detection limit. The performance of the broilers within the groups was balanced, and they did not show an abnormal behaviour or other indications due to DON contamination. The progression of growth and the feed intake were within the range of expectation (Table 3.3). No behavioural or macroscopic, pathological alterations were detected in the appearance of liver, stomach, intestine, spleen, heart, muscle and lung of the broilers receiving the mycotoxin-contaminated feed.

Table 3.3: Performance parameters

	Control	%	DON 1	%	DON 2.5	%	DON 5	%
BW day 22 [g] ± s.e.	1118 ± 36.0		1140 ± 19.8		1201 ± 30.1		1084 ± 80.4	
FI week 3 [g]	807		839		908		771	
FCR [g/g)]	1.50		1.53		1.60		1.54	
AME P1 [MJ/kg] / [%]	15.27	100	14.97	98.1	15.44	101.2	14.75	96.6
AME P2 [MJ/kg] / [%]	14.74	100	14.73	99.9	14.89	101.1	13.87	94.1
N Retention P1 [%]	67.74	100	64.72	95.5	69.26	102.3	66.47	98.1
N Retention P2 [%]	65.08	100	63.64	97.8	63.79	98.0	59.20	91.0

BW: Body weight. FI: Feed intake. FCR: Feed conversion rate (g gain / g feed). AME P1/2: Apparent metabolizable energy in MJ/kg dry matter and % relative to the control. P1/2: Period 1 or 2.

Gene expression analysis

The parameters for the RNA quality and quantity as well as the microarray quality parameters were within the requirements. The expression of a total of 566 genes was significantly altered in DON groups compared to the control group (appendix B, tables 8.1 and 8.2). The heatmap of the cluster analysis of all significantly altered genes between the control and the DON-containing groups is shown in Fig. 3.1.

Figure 3.1: Heatmap of cluster analysis

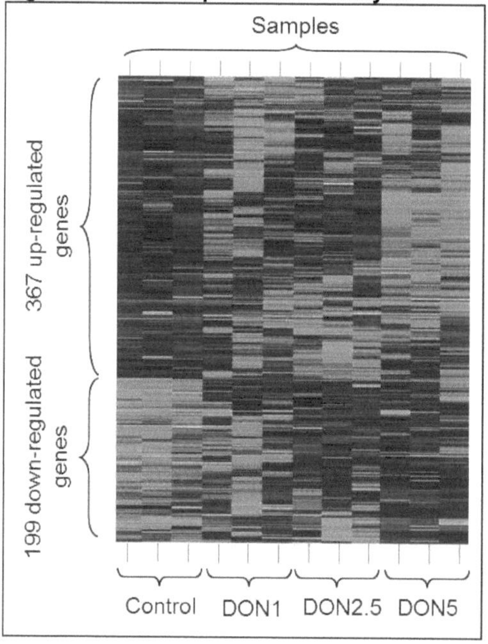

Heatmap: Green: Up-regulated genes. Black: Normal expression Red: Down-regulated genes. The regulation is displayed according to significant alteration between control group and DON-containing groups. Each horizontal line corresponds to a normalized expression of a gene from the respective sample.

The entire list of up- and downregulated genes is available in Appendix B. In the DON groups, 367 genes were upregulated. Of these, 230 genes have been specified in relation to their function or structure and for 138 genes only the transcript sequence is known. The expression of 152 upregulated genes was above the 80th percentile, which was indicated in the column "Fold change" by an asterisk (*) (table 8.1 and 8.2). The fold change of 28 genes was above 2. The Venn diagram is presented in figure 3.2 with the distribution of significantly altered genes between the treatment groups. The genes with an upregulated expression in the DON groups are presented in table 8.1.

In table 8.2, the significantly downregulated genes are indicated according to the DON groups. Of 199 significantly downregulated genes, 139 genes were known. Sixty genes were unknown or no function could be assigned to them. Fourteen genes had a fold change higher than 2 and 45 genes had a raw value above the 80th percentile.

Figure 3.2: Venn diagram of significantly altered gene expression

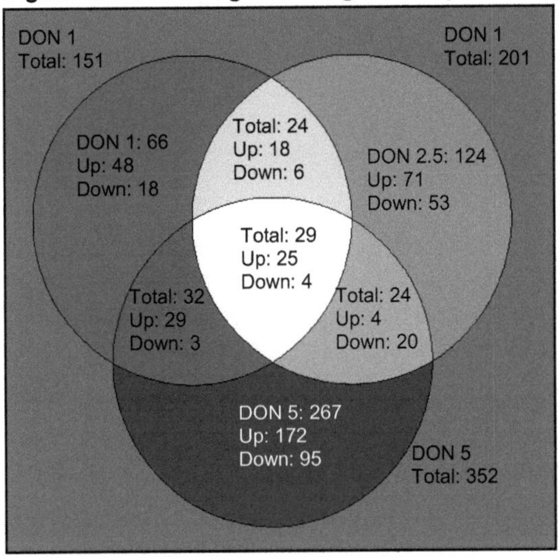

Outside of the Venn diagram the numbers of genes per group with a significantly altered expression are indicated. In the Venn diagram the numbers of total, up- or downregulated genes in the corresponding areas are annotated.

Reverse transcription and real-time PCR

The results from the microarray experiment were verified with RT qPCR, and the results are shown in Table 3.4. The alterations could be confirmed for 9 of the 15 genes in the liver. A total of 10 of 17 genes showed a significantly altered expression in the liver and the jejunum. The correlation between the fold changes of the qPCR experiment and the microarray analysis in group DON5 was 0.75. XRN1 and SLC16A1 expression were measured due to their biological relevance. The genes AKR1B1, MIA2, SLC2A5, SLC7A10 and LAPTM4B were significantly regulated in the liver, measured with qPCR and microarrays, as well as in the jejunum tissues. SLC27A4, CASP1 and EP300 had a significant altered expression in the microarray experiment in the liver and in the qPCR experiment in the jejunum. The alteration of the expression for the genes EXOSC9, TJP1, EIF2AK3 and IFT57 could be confirmed only in the liver by qPCR. SLC16A1 was not significantly altered in any measurement in the liver, but showed a decreased expression in the jejunum groups. The reduced expression of XRN1 did not reach significance in the microarray analysis ($p<0.08$), but was significantly reduced in the qPCR measurement. The expression alteration of the genes

MAPKAPK3, STK39 and TP53I3 could not be confirmed with qPCR in any tissues. Taking the liver and jejunum groups together, 14 of 17 expression alterations could be confirmed with qPCR.

Table 3.4: RT real-time PCR results

	Liver					Jejunum			
	Microarray	Don 2.5		Don 5		Don 2.5		Don 5	
Gene name	Fc	Fc	P	Fc	P	Fc	P	Fc	P
AKR1B1[1]	- 3.511	- 2 .08	**	- 2.50	*	- 1.36	NS	- 2.49	***
LAPTM4B	+ 3.058	+ 1.23	NS	+ 1.77	*	+ 1.31	*	-1.05	NS
EIF2AK3	- 1.286	- 1.64	*	- 1.70	*	- 1.18	NS	- 1.34	NS
MIA2	- 2.544	- 1.54	*	- 1.65	*	- 1.14	NS	- 1.35	*
EXOSC9	- 1.503	+ 1.36	NS	- 1.61	*	+ 1.14	NS	- 1.24	NS
TJP1	-1.520	+ 1.66	NS	- 1.66	*	+ 1.15	NS	- 1.24	NS
XRN1	NS (P<0.08)	+ 1.03	NS	- 1.25	*	- 1.02	NS	- 1.14	*
SLC7A10	+ 1.263	- 1.42	NS	+ 1.49	*	- 1.76	NS	+ 2.52	*
SLC2A5[1]	- 1.259	- 1.54	**	- 1.05	NS	- 1.72	**	- 1.94	**
SLC16A1	NS	- 1.11	NS	-1.10	NS	- 1.37	*	- 1.47	*
SLC27A4	+ 1.424	-1.43	NS	- 1.18	NS	- 1.68	NS	- 1.87	*
IFT57	+ 1.953	+ 1.78	**	+ 1.33	NS	+ 1.38	NS	+ 1.41	NS
CASP1	+ 2.198	- 1.13	NS	+ 1.04	NS	- 1.43	**	- 1.39	**
EP300[1]	+ 1.321	- 1.26	NS	- 1.60	NS	- 1.52	***	- 1.17	NS
MAPKAPK3	+ 1.299	- 1.32	NS	- 1.28	NS	- 1.15	NS	- 1.15	NS
STK39	+ 1.650	- 1.03	NS	- 1.53	NS	- 1.22	NS	- 1.01	NS
TP53I3	+ 2.646	- 1.07	NS	+ 1.06	NS	+ 1.20	NS	- 1.14	NS

n = 5 animals per treatment group.
1: Raw value above 80th percentile.
Fc: Fold change.
P: P-value of Student's t-test
Fc and P were measured between control and the corresponding DON-containing group.
*: $P<0.05$
**: $P<0.01$
***: $P<0.001$

Discussion

The use of naturally mycotoxin-contaminated wheat resulted primarily in a high DON-contamination (4.425 mg/kg) and a low calculated contamination with T-2 (5.1 µg/kg) and zearalenone (ZON, 33.7 µg/kg) in the final feed mixture of DON5. Because of the low calculated concentrations of T-2 and ZON around or below the detection level, they were measured in naturally mycotoxin-contaminated wheat. The other measured mycotoxins in the mycotoxin-contaminated wheat were below the detection limit. The maximal allowed concentration for DON in poultry feed is 5 mg/kg feed and 150 µg/kg feed for T-2. No maximal guidance values exist for ZON (bovine: 500 µg/kg feed), due to the high tolerance of poultry. The maximal permitted mycotoxin concentrations are related to the level at which no damage to the birds is expected [98]. The mycotoxin analysis in the final feed mixture showed in comparison to the maximal permitted concentration a proportion of 88.5% of DON, 3.4% of T-2 and 6.7% of ZON (bovine guidance value) in the feed of group DON5. Because of the high proportion of DON and the low amount of

T-2 and ZON, we considered the observed alterations on the transcriptional level due to effects of DON. Most of the reported effects have been investigated mainly in cell cultures or model organisms such as mice. In our experiment we showed for the first time similar alterations in the gene expression in broilers.

The growth, feed intake and feed conversion rates were within the normal range in our experiment. Although body weight and feed intake were numerically the lowest in group DON5 and the highest in group DON2.5, this was statistically not significant. It has been mentioned by Awad *et al.* (2008) [12] that moderately high concentration levels of DON can even have growth-promoting effects. Further, the increased standard error in body weight of group DON5 indicates that different sensitivity toward DON-contaminated feed might exist, depending on the genetic background of the animals.

Nutrient transport

The gene expression analysis has shown that the nutrient uptake into the cell seems to be influenced by the mycotoxin-contaminated feed. The passive D-fructose transporter with a very low glucose affinity, SLC2A5 (fc:-1.259, DON 2.5), had a decreased expression in the liver (fc:-1.54, DON2.5) and the jejunum (fc:-1.94, DON2.5/5). The numerical decrease in AME by 5.9% might be a further indication of a reduced monosaccharide uptake in the jejunum. Maresca *et al.* (2002) [99] showed in their study that fructose uptake into the cell and corresponding transporter SLC2A5 were inhibited by 42% in the intestinal cell line HT-29-D4 cells due to DON administration. Our findings indicate that similar effects occur in the jejunum of broilers in the living animal and to a lesser extent in the liver. The gene *SLC27A4* (fc:+1.427, DON5) is a palmitate transporter and the increased expression could not be confirmed with qPCR in the liver. But the expression was significantly reduced in the jejunum (fc:-1.87, DON5). In primary enterocytes treated with *SLC27A4* antisense, palmitate and oleate uptake were reduced simultaneously [100]. Maresca *et al.* (2002) [99] reported an increased palmitate uptake by 35% in their study after challenging HT-29-D4 cells with DON. We assume that chronic exposure to DON can invert the regulation or that another function mechanism on the transport activity is responsible for the discrepancy between the decreased expression of *SLC27A4* in the jejunum and the increased palmitate uptake reported by Maresca *et al.* (2002) [99]. The SLC7A10 protein (fc:+1.26, DON5) transports D-serine, glycine and L-isomers of alanine, serine, threonine and cysteine [101]. The expression change was confirmed with qPCR in the liver (fc:+1.49, DON5) and the jejunum (fc:+2.52, DON5). Further, the N retention was numerically reduced by 9%, which gives an indication on the amino acid uptake. The decrease in L-serine uptake in the cell culture study of Maresca *et al.* (2002) [99] is in disagreement with our observation that *SLC7A10* showed an up-regulated expression, but is in agreement with the reduced N retention in group DON5. The discrepancy could indicate a direct inhibition of this L-serine transporter by DON, resulting in the up-regulation of SLC7A10. SLC7A5 (fc:-1.218, DON5) is a transporter of neutral amino acids with branched or aromatic side chains like leucine [102]. In an incorporation study with mice the ^{14}C leucine incorporation into the liver was decreased at 20mg DON/kg feed. It was assumed that DON inhibited the protein synthesis and therefore the reduction of the ^{14}C leucine amount occurred [103]. The addition of 0.5 and 5 µg DON/ml to Caco-2 cells led to a decreased [^{3}H] leucine incorporation and a parallel decrease of cellular protein content [104]. To what extent the leucine uptake into the cell is responsible for the decrease of the protein synthesis stays unclear. In a related, unpublished study *SLC16A1* expression was significantly altered and therefore the test was also used to verify the expression in the jejunum in the current study. *SLC16A1* (fc:-1.87, DON5) had a decreased expression in the jejunum, and it transports monocarboxylates like butyrate, lactate and pyruvate [105]. It has been suggested that effects of DON on membrane transport is a result of specific modulation of the transporters and is not due to cell damage [99], which is supported by our observations, especially because *SLC7A10* was up-regulated. The alteration of these five gene expression levels influence the nutrient uptake into the cell and might have implications for the growth rate of birds.

Detoxification

AKR1B1 (fc:-3.511, DON2.5 and 5) had a decreased expression in the liver, measured with microarray and qPCR experiments (fc:-2.5, DON2.5 and 5), as well as in the jejunum (fc:-2.49, DON5). AKR1B1 reduces reactive aldehydes and ketones to primary and secondary alcohols. Reactive oxygen species also target polyunsaturated fatty acids that form lipid hydroperoxides, which are finally decomposed to form reactive lipid aldehydes, which are highly mutagenic. AKR1B1 detoxifies those lipid aldehydes, which can react with bases in DNA to form etheno- and heptano-etheno-DNA adducts [106]. The down-regulation of *AKR1B1* by chronic DON exposure might be also in relation to several up-regulated DNA repair genes, which are discussed in the section below. In the human hepatoma-derived cell line HepG2, the expression of *AKR1B1* was significantly reduced after the cells were challenged with a series of cytotoxic substances [107]. In conclusion, *AKR1B1* seems to be an interesting, *in vivo* biomarker for DON-induced alteration of gene expression.

DNA repair

Six genes were significantly up-regulated in DON5 and one in DON2.5, which are responsible for DNA damage repair and recognition, showing a possible effect of DON on DNA integrity. The six genes correspond to 4.1% of significantly altered genes in a DON5 group with a known function (124 genes). The protein PARP1 (fc:+1.410*, DON5) is involved in DNA repair and has a critical role in signalling DNA single-strand breaks [108]. Shifrin and Anderson [109] showed that the addition of DON to the Jurkat human lymphoid cell line led to the activation of pro-caspase-3, leading to a higher cleavage of PARP1, a substrate of caspase-3. In murine macrophages J774A.1, an increased amount of PARP1 and significantly increased amount of cleaved PARP1 protein resulted after the challenge with DON [110]. The up-regulation of PARP1 therefore shows that similar effects might occur in the liver of broilers. MPG (fc:+1.406, DON5) is responsible for the recognition and excision of alkylated purine bases [111]. EME1 (fc:+1.366, DON5) cleaves branched DNA structures. In the absence of EME1, chromosomal aberrations occurred and led to genomic instability [112]. The gene RAD51L1 (fc:+1.395, DON2.5) functions in homologous recombination and DNA double-strand break repair [113]. CHAF1B (fc:+ 1.435, DON5) is part of the complex CAF-1 and the only subunit available on the Chicken Genome Array. The complex is related to nucleotide excision repair and to the repair of single-strand breaks [114]. RIF1 (fc:+1.235, DON5) functions as a DNA damage response factor [115]. XPAC (fc:+1.279, DON5) is involved in nucleotide excision repair [116]. In male broiler chickens fed with 10 mg DON/kg feed, a comet assay was performed in spleen leukocytes, which detected aberrations, and a significantly higher amount of DNA breaks were found [117]. Interestingly, all the DNA repair-related genes mentioned above were only up-regulated in group DON5 with the exception of *RAD51L1*, which was up-regulated in DON2.5. Genes like *PARP1*, *MPG*, *EME1*, *XPAC* and *CHAF1B* belong either to the base excision repair or to the nucleotide excision repair pathway. Under reduced activity of AKR1B1, the anchorage of lipid aldehydes to the DNA might have occurred with a higher frequency, and therefore the discussed single-strand repair genes might have shown a higher expression in the group DON5. A single-strand break or modified nucleotides do not lead necessarily to double-strand breaks or, as a result, to detection in the comet assay. Continuing, it seems that the activation of DNA damage-related genes is induced only by a concentration of 4.42 mg DON/kg feed.

Translation initiation

The expression decrease of *EIF2AK3* (fc:-1.286, DON2.5/5) could be confirmed in the liver with qPCR (fc:-1.70). EIF2AK3 is a target for transcriptional up-regulation of the unfolded protein response and phosphorylates eIF2α, a translation initiator [118]. The phosphorylation of eIF2α inhibits translation initiation and protein synthesis. EIF2AK3 activation and the subsequent eIF2α phosphorylation prevent the cell from the accumulation of incorrectly folded or unfolded proteins [119]. *DNAJC3* (fc:-1.435, DON2.5) is induced during endoplasmic reticulum (ER) stress [120]. DNAJC3 binds to EIF2AK3 and disrupts the activity of EIF2AK3. Subsequently, eIF2α phosphorylation by EIF2AK3 is reduced, stimulating the eIF2α-induced protein synthesis [121]. Our results are consistent with the publication of Yang *et al.* (2000) [118], which showed reduced expression level of *DNAJC3* after 24 hours in murine thymoma cell line EL-4 treated with DON. The binding of deoxynivalenol to the 60S ribosomal subunit and the subsequent translation inhibition might be counterbalanced by the down-regulation of *EIF2AK3* and *DNAJC3*. Pestka [122] assumed that EIF2AK3 might have an important role in the signalling after the challenge with DON, due to influenced up- and downstream signal transducers. The down-regulation of *EIF2AK3* in our experiment confirms this assumption. The ribotoxic stress response seems to have some pathway elements similar to the ER stress response, but with inverse gene expression regulation.

Stabilization of mRNA

The expression change of *EXOSC9* (fc: -1.503, DON5) was confirmed with qPCR (fc:-1.61) in the liver. EXOSC9 is a subunit of the exosome, a 3'-5' exoribonuclease complex. This complex processes ribosomal and small nuclear RNAs and degrades mRNAs. In HeLa extracts the addition of RNA containing an adenylate/uridylated (AU)-rich element (ARE) increased their 3'-5' exonucleolytic decay rate. The inactivation of EXOSC9 highly decreased the efficiency of the exonucleolytic decay, showing the central role of EXOSC9 in mRNA decay [123]. In human HEp-2 cells, it has been shown that the knockdown of EXOSC9 led to a twofold increase in *β-globulin* containing an ARE [124]. The 5'-3' exoribonuclease *XRN1* (fc:-1.796, DON5) was numerically down-regulated in the microarray experiment and significantly down-regulated in qPCR experiment (fc:-1.25) in the liver. XRN1 is a component of the alternative 5'-3' mRNA decay pathway [125]. The knock-down of XRN1 led to a 3.5-fold increase of *the β-globulin-ARE* mRNA. Therefore, EXOSC9 and XRN1 seem to be involved in the turnover of ARE-containing transcripts [124]. In his review, Pestka [126] summarised the genes with a lower mRNA decay rate upon DON administration, which are *COX-2*, *TNF-α* and *IL-6* in macrophages and *IL-2* in EL-4T-cells. It has been reported that all four mRNAs are stabilised by ARE in the 3' untranslated region [127-129]. The down-regulation of *EXOSC9* and *XRN1* might be the reason for increased mRNA stability of several ARE-containing genes in DON-challenged cell cultures. Further, the reduced mRNA depletion, due to *XRN1* and *EXOSC9* down-regulation, might have also led to higher mRNA levels in general. In the final gene list (supplementary files) a disequilibrium exists with 367 up-regulated genes and 199 down-regulated genes. The possible higher mRNA amount might compensate for the possibly reduced translation, due to binding to the 60S ribosomal subunit by DON. *EXOSC9* and *XRN1* seem to be interesting biomarkers for mRNA stabilisation and the decay rate of ARE-containing mRNA.

Tight junctions

Cells are connected by intercellular complexes like tight junction and adherens junctions. In Caco-2 cells it has been shown that the transepithelial electrical resistance (TEER) decreased to 30% after 24 hours of DON incubation at a concentration of 2 μg/ml. It was mentioned that TEER reflects the functional tight junctions made of transmembrane proteins [130]. Amongst the proteins mentioned was TJP1 (fc:-1.52, DON5), which was down-regulated in our experiment in the liver, which was also measured with qPCR (fc:-1.66). In porcine IPEC-1 and -J2 cells the treatment with 2 μg DON/ml let to disintegrated TJP1 structure and the total amount of TJP1 moderately decreased, measured with Western blot [131]. It was reported that TJP1 interacts with CLDN3 (fc:+1.637, DON1/5) [132]. Pinton *et al.* (2009) [133] have shown a decreased CLDN3 protein amount in porcine intestinal epithelial cells (IPEC-1) after the treatment with DON for 48h with a parallel decrease of TEER. Therefore the up-regulation of *CLDN3* might counterbalance the reduction caused by DON over a feedback mechanism. The decrease of TEER shows the acute effects in cell cultures. The extent to which the chronic administration of DON leads to the same effects has not been reported. The significant expression change of TJP1 and CLDN3 indicates that similar effects, such as the decreased TEER in cell cultures, might also occur in the liver of living animals.

Immune-related genes

The expression of *MIA2* (fc:-2.544, DON5) was significantly decreased, measured with qPCR in the liver (fc:-1.70) and the jejunum (fc:-1.35). *MIA2* expression is regulated by IL-6 and TGF-β, which increased the expression level of *MIA2* in human hepatocytes and in HepG2 cells [134]. The down-regulation of *MIA2* in the liver and the jejunum may indicate a reduced amount of IL-6 in the liver cells of broilers. In murine macrophages it has been shown that IL1RL1 (fc:+1.661, DON2.5/5) exerted anti-inflammatory effects by the negative regulation of the IL-6 production [135], which affirmed the previous results with the decreased MIA2 expression. *In vitro* experiments with the addition of DON exhibited that the half-life of the ARE-containing mRNA IL-6 and TNF-α and the amount of the respective proteins were increased within 24 hours [136]. IL-6 seems to be critical for regulation of IgA. In broilers fed with 7 to 9.2 mg DON/kg feed, the biliary IgA concentration was decreased [137]. Azcona-Olivera *et al.* (1995) [138] showed in mice exposed to a 25 mg DON/kg diet an increased expression of *IL-1β, IL-6* and *TNF-α* mRNA. Four hours after the administration of DON, the cytokines were again only slightly up-regulated in the liver [138]. This may also explain the findings that *IL-1β, IL-6* and *TNF-α* mRNA were unchanged in our experiment. IL1RL1 might be an antagonist to the deoxynivalenol by stabilising IL-6 and TNF-α levels. Especially in long-term administration of deoxynivalenol, the normalisation of the mRNA and cytokine levels might be economical. The up-regulation of pro- and anti-inflammatory factors may influence the sensitivity of the immune system and act immune-stimulating. *LAPTM4B* (+3.058, DON1/5) expression increase could be confirmed in the liver (fc:+1.77, DON5) and the jejunum (fc:+1.31, DON2.5). LAPTM4B is a tetra-transmembrane glycoprotein localised on endosomes and lysosomes [139]. LAPTM4B was also up-regulated in the early phase of liver regeneration in rats [140]. Therefore, LAPTM4B might be used as a biomarker for liver regeneration.

Complex I-related proteins (not part of the accepted publication)

The NADH dehydrogenase Fe-S protein 1 (NDUFS1, fc: +1.802, DON5) is part of the complex I. Complex I is a multiprotein complex located in the inner mitochondrial membrane. Complex I functions as transporter of electrons from NADH to ubiquinone, leading to the translocation of protons from the mitochondrial matrix. It has been shown that the cleavage of NDUFS1 by caspases during apoptosis is responsible for disruption of the electron transport and the mitochondrial transmembrane potential and subsequently the production of ROS, loss of ATP production and mitochondrial damage [141]. The NADH dehydrogenase 1 beta subcomplex, 1 (NDUFB1, fc: +1.342, DON5) and NADH dehydrogenase 1 alpha subcomplex, 7 (NDUFA7, fc: + 1.841, DON1) are also a subunit of complex I [142]. It seems that complex I is directly related to the process of apoptosis and at least one (NDUFS1) is directly targeted by the apoptotic protein caspase-3. The upregulation of the three genes could have led to reduced sensitivity to apoptosis as a result of long-term administration of DON.

Miscellaneous (not part of the accepted publication)

Kruppel-like factor 11 (KLF11, fc: +2.09, DON1/2.5) is induced by TGF-β (unchanged) and is a transcription factor. In OLI-neu cell lines the treansient overexpression of KLF11 resulted in elevated apoptosis. KLF11-transfected cell let to a decrease of Bcl-XL (unchanged) protein amount and the induction of the caspase-3-depentent apoptosis. Bcl-XL is an anti-apoptotic enzyme by preventing cytochrome c (unchanged) release [143]. In RAW 264.7 cell lines DON induced transcription of TGF-β and other cytokines [144]. The upregulation of KLF11 might indicate an increase in TGF-β signaling.

Little is known about the BTB (POZ) domain containing 7 (BTBD7, fc: -1.77, DON5). Cell lines were transfected with the BTBD7 gene and injected subcutaneously in nude mice BEL7404. The overexpression of BTPD7 led to tumor growth within 3 weeks. A subsequent injection of cells containing antisense sequence reduced the tumor weight to 40% of the control [145]. The downregulation of BTBD7 might have anti-proliferative effect on liver growth and function as a tumor suppressor. The glutamic-oxaloacetic transaminase 1 (GOT1, fc: +1.31, DON5) which is also called aspartate aminotransferase 1. In the plasma of boiler chickens aspartate aminotransferase activity was numerically increased due to DON feeding at the level of 3 mg/kg feed for 42 days [146].

The protein S (alpha) (PROS1, fc: + 1.81, DON5) was upregulated between DON 0/1 and DON 2.5/5 and the raw expression values were above the 80th percentile. In HepG2 cell cultures the administration of IL-6 led to an expression increase of PROS1. PROS1 contains a STAT3 and a C/EBPβ binding site in the promoter region. Western blot analysis showed an increase of phosphorylated STAT3 but not of C/EBPβ. It has been discussed that the upregulation of PROS1 might be related to the cell survival in inflamed region [147].

AT rich interactive domain 1A (SWI-like) (ARID1A, fc: -1.66, DON1/2.5/5) was downregulated in all DON-containing groups and is a homologue of yeast SWI 1. It contains a DNA-binding motif (ARID) and is a member of the SWI/SNF complex, an ATP-dependant chromatin-remodeling multiple subunit enzyme. ARID1A seems to be a tumor suppressor gene [148]. According to Wang (2004) human tumor cell lines showed a reduced expression of ARID1A and also a reduced protein level. SWI/SNF complexes facilitate the binding of transcriptional regulators and factors to the DNA. They also control differentiation and development and are critical for negative control of cell proliferation. ARID1A and ARID1B (fc: - 1.407, DON2.5) are closely related proteins and alternative subunits of the SWI/SNF complexes. In cell models, ARID1A-containing SWI/SNF complexes had an anti-proliferative function whereas ARID1B-containing SWI/SNF complexes had a pro-proliferative function [149].

p53 related change of gene expression (not part of the accepted publication)

DON induced the phosphorylation and activity of p53 in leukocytes [150]. Several genes were found in our experiment, which are related to transcription factor p53. The aldehyde dehydrogenase 4 family, member A1 (ALDH4A1, fc: + 1.33, DON5) is a mitochondrial matrix protein that performs the second step of the proline degradation pathway. ALDH4A1 was found over the screening for p53-induced genes in response to DNA damage and adriamycin treatment. The viability of cells overexpressing ALDH4A1 was increased in p53-null cells. It has been proposed that ALDH4A1 inhibits H_2O_2-induced generation of ROS and the subsequent apoptosis [151]. The ubiquitin-conjugating enzyme E2D 1 (UBE2D1, fc: + 1.26, DON2.5) seems to function as an E2 ubiquitin-conjugating enzyme which is involved in the p53 degradation [152].
The inhibitor of growth family, member 3 (ING3, fc: +1.214, DON5) activates the p53-responsive promoters of p21/wafl and Bax genes. ING seems to modulate p53 function as a transactivator, which suppresses the colony formation efficiency and induces apoptosis in cell cultures [153]. Zhou *et al.* (2005) mentioned that p53 can upregulate Bax transcription due to DON administration [150]. ING3 might be linkage between p53 activation and upregulation of Bax. In HepG2 cell lines ING3 was also upregulated after phenol or antimony (Sb) administration [154]. N-myc downstream regulated gene 1 (NDRG1, fc: -1.40, DON) is a p53-regulated gene and its expression is induced by DNA damage. Further it is related to caspase-3 activation [155]. It has been shown that activated p53 induced upregulation of the expression of STEAP family member 3 (STEAP3, fc: -1.30, DON5) in the p53-inducible cell line LTR6. Overexpression of STEAP3 led to increased apoptosis sensitivity. STEAP3 antisense-containing LTR6 cells were generated and p53 was activated. 18 hours after the p53 activation, the cleavage of ADPRT (upregulated) was diminished and also a significant reduction in cell death was observed [156]. The p53-induced CASP3 cleaves ADPRT and leads to apoptosis [157]. The function of STEAP3 seems to be ambivalent. The retinoblastoma binding protein 6 (RBBP6, fc: -1.65, DON5) binds to p53 and to Rb1 both tumor suppressors. Scott *et al.* (2005) reported that high expression of RBBP6 of mice was associated with decreased expression of 29 ribosomal proteins [158]. In our study only mitochondrial ribosomal protein L46 (fc: +1.26) was upregulated in group DON 5. This could be in relation of the low DON doses used or the low change of expression of the ribosomal subunits. It has also been reported that DON inhibits the 60S ribosomal subunit and subsequently the translation [159]. RBBP6 seems to be a biomarker for translation inhibition over the ribosomal subunits. It seems that p53 is also activated due to long-term administration of DON in broiler liver or has an influence in DON-induced transcription regulation. The transcription regulation is often performed by several transcription factors and the importance of a single transcription factor on the expression level of a gene seems to be variable [160]. A conclusion from an altered gene expression on a transcription factor can be misleading, but might give an indication on signaling pathways of activated transcription factors.

Experimental background (not part of the accepted publication)

The parameters for the RNA quality and quantity as well as the microarray quality parameters were within the requirements. Elevated variances in the gene expression profile between the birds within the same group were observed and are presumably due to the use of commercial broilers. One reason could be that the broilers were F2 hybrids with high heterozygosis compared for instance to inbred laboratory mice with high genetical homozygosis. In poultry production, different roosters are used for breeding. Even slightly uneven genetical material from the paternal side and also from the maternal side might lead to elevated variance in the gene expression profile. Therefore the differences in expression fold change were lower compared to other studies [161, 162]. On the other hand, the standard error in group DON5 was increased and may explain a part of the variability seen on the level of the gene expression.
In the microarray analysis the normalization was performed on the entire signal of the chip as well as every sample was divided by its median in all chips. Therefore the normalization has a higher precision than the real-time PCR experiments, which was only normalized to two genes. The endogenous controls were selected according to microarray data and they had to have a high expression above the 90th percentile, a similar expression between the groups and a low variability within the group. The two endogenous controls *MRPS18A* and *RPS17* turned out to perform the best compared to the traditional and other selected control genes such as: *GAPDH*, *β-actin*, *DHRS7*, *RPS28*, *APOB* and *LDHA*.

Conclusion

We could confirm several results in the living animal from previous studies, which were performed mainly in cell cultures. In addition new genes with an altered expression upon DON supplementation were detected and this may be helpful for further investigations. The main effects with economic impact might be the altered expression of the solute carrier transcripts, which transport D-glucose/D-fructose, palmitic acid, monocarboxylates, L-serine, leucine and other amino acids. In the microarray analysis, the candidate genes *SLC2A5*, *SLC27A4*, *SLC7A10* and *SLC7A5* were significantly altered and are in agreement with the observations of other authors. Further, the translation initiation-related genes *EIF2AK3* and *DNAJC*, as well as the mRNA stabilization genes, could have an important influence on the protein synthesis rate and on the growth of broilers. The upregulation of DNA repair proteins give an indication of possible mutagenic effects of DON or its decomposition products. The detoxification enzyme AKR1B1 is an indicator that at 2.5 mg DON/kg feed, the concentration might be high enough to show cytotoxic effects. In consequence of the cytotoxic effects the expression of DNA repair proteins is increased. The verification of the results in the jejunum from the microarray and the real-time PCR analysis in the liver increases the reliability of the DON-regulated genes. In general, effects on the level of RNA expression cannot be directly transposed from one organ to others. Therefore the parallel alterations in the liver and the jejunum might indicate a general function mechanism of DON on the different relative amounts of transcripts. The maximum allowed DON contamination in poultry feed in the European Union is 5 mg DON/kg feed [88]. At a concentration of 2.5 mg DON/kg feed significantly altered gene expression was observed. The results of this study also show that the gene expression analysis is a very sensitive method to detect alterations and is a valuable starting point for further systematical investigations, especially for substances, whose effects have not yet been studied intensively.

4. Impact of highly mycotoxin contaminated feed on the gene expression in the liver of broilers

Abstract

Mycotoxin deoxynivalenol (DON) is a secondary metabolite from the *Fusarium* species and is frequently present in wheat and other cereals. The effects of DON intake range from a reduction in feed consumption, weight gain, and nutrition efficiency to the suppression of the immune system. DON binds to the 60S ribosomal subunit and inhibits subsequent protein synthesis at the translational level. It has been suggested that cells and tissues with a high protein turnover rate, like the liver, are most affected.

Twenty one-day-old broilers per group were fed the following diets for 35 days: control, contaminated wheat (D20), and contaminated wheat + 0.2% polymeric glucomannan mycotoxin absorber (MD20). The main contaminant was DON (17.7mg/kg), with lesser amounts of zearalenone (0.14mg/kg). T-2 and HT-2 were below the detection limit. Gene expression was determined in the livers of seven animals in the control and group MD20 and eight animals in group D20. Genes selected for the analysis had to have a fold change (fc) of 1.2 and a p-value below 0.05. Glucose, total cholesterol, and low-density lipoprotein (LDL) cholesterol concentrations were determined in heparinized blood plasma.

At the end of weeks two and three, the body weight was significantly reduced in D20; however, the mycotoxin absorber prevented this effect in MD20. The birds from D20 recovered from their growth depression in the last two weeks of the feeding trial and showed only a numerically reduced body weight at the end of the experiment. Feed consumption was significantly reduced in D20 at the end of week three. During the experiment, total feed consumption was significantly reduced in D20 and MD20. By group, 195 genes were significantly regulated in D20, 433 genes in MD20, and 635 genes in the comparison between D20 and MD20. In D20, the genes were mainly related to increased xenobiotic metabolism and the prevention of the cell cycle arrest in the G_2/M phase. The increased expression of cell-cycle-related genes and the significant increase of the apparent metabolizable energy indicated that the development of the birds was delayed. In MD20, changes in expression were observed for cell permeability, hormone signaling, cholesterol biosynthesis, and LDL-related genes. LDL-cholesterol concentration in the MD20 blood plasma increased significantly and seemed to be caused by the decreased expression of the *LDL-receptor* (fc: -1.70) and the increased expression of the LDL-shell protein *APOB* (fc: 1.24). The total cholesterol concentration in the blood plasma was similar in all groups, even though 13 of 23 genes of the cholesterol biosynthesis pathway were downregulated. The immune system seemed to be influenced by the downregulation of *SOCS2* (fc: -1.48, D20 and -1.72, MD20) and other related genes. In comparing D20 to MD20, the insulin signaling cascade was deregulated and the gluconeogenesis-related genes were upregulated. The blood plasma glucose concentration was not significantly altered, although the numerical decrease between D20 and MD20 corresponded to the expression change of the microarray analysis.

The mycotoxin absorber could protect the animals from detrimental effects; nevertheless, the expression changes of several genes clearly indicated the influence of DON and its incomplete immobilization.

Introduction

Deoxynivalenol (DON) is a secondary metabolite from the *Fusarium* species (*F. graminearum, F. culmorum*) and belongs to the group of trichothecenes, which are structurally related. About 25% of the world's food crops are contaminated with mycotoxins, which are, in general, fungal-produced toxins. Even though DON is not as toxic as other mycotoxins, it is one of the most common trichothecenes [163]. Acute exposure to high doses of DON can lead to diarrhea, vomiting, leukocytosis, and hemorrhaging, and extremely high doses can cause death. Chronic exposure to DON can reduce feed intake, weight gain, and nutritional efficiency. Depending on the dose, frequency, and duration, DON exposure can lead to stimulation or suppression of the immune system [164]. Low doses of DON increase the expression of cytokines and chemokines. High doses, in contrast, lead to the rapid apoptosis of leukocytes [13]. Decreased feed intake and weight gain have been observed in pigs fed 3 mg/kg DON either in a purified form or in the form of naturally contaminated corn at day 7. Until the end of the experiment (day 32), the animals receiving the purified DON recovered from the growth depression in contrast to the animals receiving naturally contaminated corn. The differences between naturally DON-contaminated feed sources and the feed spiked with purified DON might be caused by other myctoxins (acetyl-DON, DON glucosides) or fungal components present in the naturally contaminated feed [165]. We decided to use naturally contaminated wheat for the experiment to reproduce the effects that occur in the commercial broiler production.

The reduced weight gain is caused partially by the reduced feed intake, which might be influenced by an increased serotonin concentration [13, 166]. On the molecular level, DON interferes with the 28S ribosomal RNA and promotes cleavage in the peptidyltransferase site, thereby inhibiting the translation and the protein synthesis [167]. This leads to the rapid activation of mitogen-activated protein kinases (MAPK) via a mechanism known as "ribotoxic stress response." The main signal transducers, activated by the DON administration, are extracellular signal-regulated protein kinase 1 and 2 (ERK1 and 2), c-Jun N-terminal kinase 1 and 2 (JNK1/2), and MAPK14. Beside the signal transducers several transcription factors are activated or increased by DON, like jun proto oncogene (c-Jun), early growth response 1 (EGR1), cAMP responsive element binding protein 1 (CREB1), and CCAAT/enhancer binding protein beta (CEBPB), which are critical for the regulation of immune- and inflammation-related genes. The administration of DON *in vivo* can cause apoptosis in bone marrow, Peyer's patches, thymus, and leukocyte [13]. Beside the activation of the apoptotic pathway, DON also activates the competing survival pathway over the AKT and the downstream substrate FOXO1 phosphorylation [150]. In addition to the apoptotic pathway, DON causes cell cycle arrest in the G_2/M phase over the activation of MAPK14 and TP53, leading subsequently to the increased expression of *p21 in vitro*. p21 is a critical inhibitor of the cyclin-dependent kinases, which modulate the cell cycle at various levels [168, 169].

The intensive *in vitro* research brought up the major signaling mechanisms upon DON administration [13]. To what extend the cell culture results can be assigned to the effects in the target tissue of an animal is unclear. Therefore, the expression changes of up to 28,000 genes were measured in the livers of broilers to screen for the overall effects. Chicken have a high tolerance to DON-contaminated feed in contrast to other farm animals. The insensitivity to DON might be due to a low absorption rate, leading to a low bioavailability and a rapid elimination [163]. Therefore, feed producers favor to use low contaminated feed and food batches for the chicken feed production and not for sensitive animals like pigs [12]. In addition DON undergoes de-epoxidation by the gut flora, and in the liver DON seems to be conjugated to glucuronides [170]. Broilers were used because of their high growth rate and the regular contamination of their feed in commercial production. The liver is one of the first targets of DON with possible toxicological effects, due to its high protein turnover. The liver is also the main organ for detoxification and the most important organ for the metabolism. DON disappears between the crop and the jejunum from the gastrointestinal tract and high radioactivity has been observed in the liver and the bile, with the expectation of an excretion with bile back into the small intestine [12]. These reasons led to the decision to use the liver for the microarray analysis. According to the literature, DON has to exceed a concentration of 5mg/kg to cause detrimental effects on the performance parameters [12]. Several DON-related effects on the level of the gene expression were observed in our previous experiment with up to 5mg/kg DON. In the European Union, the maximal value for feed contamination are 5mg/kg DON in poultry nutrition [88]. In general, concentrations below 15mg/kg DON had no adverse effect on body weight gain, feed consumption, or feed efficiency [12]. In the previous experiment, the broilers were fed up to 5mg/kg DON in the form of naturally contaminated wheat for 23 days. Due to the low number of significantly altered genes and a low fold change among the groups, the concentration of contaminated wheat was increased to attain a level of 20mg/kg DON in the feed.

It has been reported that the inclusion of a polymeric glucomannan mycotoxin absorbent has some beneficial effects in preventing the detrimental effects of the mycotoxins [171, 172]. The mycotoxin absorbent, which is extracted from the cell wall of yeast (Mycosorb®, Alltech Inc.), should bind the mycotoxin in the lumen of the gastrointestinal tract and prevent its uptake [173].

The focus of this work was, on one side, to describe the molecular effects of DON and the mycotoxin absorber on the level of gene expression and the comparison to the *in vitro* results from previous studies. On the other side, the mycotoxin absorber should be evaluated in relation to the performance parameters of broiler chickens.

Methods

Experimental design, birds, and diets

Sixty one-day-old male ROS 308 broiler chicks were obtained from the commercial hatchery Erb Brüterei AG (Oberdiessbach, Switzerland). The birds were weighed at the beginning of the experiment and separated into three treatment groups with the same total bird weight. After a week, each initial group was divided into four subgroups, due to limited cage size. The birds were distributed to have the same average weight in all four subgroups. Finally, 12 groups were attained with five birds per cage. The broilers were held for 35 days.

The standard diet (table 4.1) was formulated according to the National Research Council (1994) and prepared for all three groups using the same process. Celite, a nutritionally inert substance, was added as an indigestible marker to the feed to increase the level of acid insoluble ash (AIA). After the basic feed mixture was prepared, 41.33 kg of DON-contaminated wheat (48.4 mg DON/kg) per 100 kg feed were added to the diet of the two DON-containing groups. Also, 41.33 kg of uncontaminated wheat per 100 kg were added to the control group. At the end, 200 g (0.2%) of yeast-based cell-wall extract (Mycosorb®, Alltech Inc., Lexington, Kentucky, USA) were added to group MD20.

According to the calculations, the control group did not contain DON, the second group contained 20 mg DON/kg feed (group D20), and the third group also contained 20 mg DON/kg feed and the mycotoxin-absorber (group MD20). Feed samples were taken from the principle components, the control and the treatment diet after pelleting and on days 1, 17, 31, and 35. All the feed samples were stored in a 4°C cooler. The pelleted feed and water were provided *ad libitum*. The total cage weight and feed intake were monitored weekly. Total water intake per cage was recorded for four days a week. Excreta samples were taken from day 15 to day 17 and from day 29 to day 31 from all 12 cages. After each sampling, they were frozen at -20°C. Dry matter, ash, acid insoluble ash, gross energy, crude fat, and nitrogen content were determined for all of the feed and excreta samples.

The DON concentrations of the contaminated wheat and the three treatment diets were determined with LC-MS/MS. T-2 toxin and zearalenone concentrations of the diets were determined with LC-MS/MS, and ochratoxin A, fumonisin B1, B2, and aflatoxin B1, B2, G1, G2 were determined with HPLC-FLD in an external laboratory (UFAG, Sursee, Switzerland). Apparent metabolizable energy and N retention were calculated according to Scott and Hall (1998) [174]. The animal experiment was performed according to the regulations of the cantonal veterinary office in Zurich. After 35 days, the broilers were weighed individually and slaughtered by cervical dislocation and bleeding. Blood samples were collected in heparinized blood collection tubes (Venoject, Terumo Europe Wettingen, Switzerland), cooled, and centrifuged at 1,000 g for 20 minutes. Then the supernatant was removed and stored at -80°C for the glucose, cholesterol, and LDL-cholesterol analyses. Liver samples were taken immediately after the blood samples. From the middle part of the left liver lobe, a slice of approximately one cm was cut and frozen in liquid nitrogen.

Table 4.1: Composition and nutrient content

	Control	D20	MD20
Calculated parameters			
Deoxynivalenol [mg/kg]	0	20	20
Yeast extract [g/kg]	0	0	2
Ingredients %			
Corn		17.7	
Soybean meal 48%		25	
Potato protein		2.5	
Rapeseed oil		5	
Salt		0.15	
DL-methionine		0.26	
Limestone grit		0.9	
Lysine-HCL		0.17	
DCP 38/40		1.2	
Sodium bicarbonate		0.3	
Celite 545		1.52	
Premix[1]		0.3	
Contaminated wheat	0	41.33	41.33
Uncont. wheat	45	3.67	3.47
Yeast extract	0	0	0.2
Analyzed parameters			
DON mg/kg	0.32	17.7	17.7
ZON mg/kg	0.01	0.14	0.14
Dry matter (%)	88.5	89.5	89.5
Gross energy (MJ/kg)	18.9	18.9	19.0
Crude fat (%)	7.2	7.2	7.2
Crude protein (%)	23.4	24.6	24.7
Ash (%)	6.85	7.14	7.12

[1]Premix: Diet according to Aeschbacher *et al.* (2005) [22]
ZON: Zearalenone
Ochratoxin A, fumonisin B1, B2, aflatoxin B1, B2, G1, G2, T-2 and HT-2 below detection limit.
Contaminated wheat (48.4 mg DON/kg); uncontaminated wheat (0.58 mg DON/kg)

High-density oligonucleotide array hybridization

For the RNA extraction, seven animals in the control and the MD20 groups and eight animals from the D20 group were selected according to the average weight (meaning the heaviest and lightest birds were excluded). Total RNA was isolated from 22 frozen liver samples with TRIZOL reagent (Invitrogen, Basel, Switzerland). The samples were homogenized in a mortar filled with liquid nitrogen. After isolation, RNA concentration and purity were determined with an ND-1000 (NanoDrop Technologies, Wilmington, DE, USA). Another 95 µg of RNA was cleaned up with the RNeasy Mini Kit with a DNA digestion step (Qiagen, Hilden, Germany). Before the reverse transcription, the RNA integrity was determined with an Agilent RNA 6000 Nano Kit (Bioanalyzer 2100, Agilent Technologies, Basel, Switzerland). The integrity number was calculated using the Bioanalyzer 2100 software with a value between 1 (for totally degraded RNA) and 10 (for undegraded RNA). For the microarray experiments, samples were only taken with a value higher than 9. Labeled cRNA probes were generated according to the protocol from Affymetrix using the GeneChip® One-Cycle cDNA Synthesis Kit (Affymetrix, Santa Clara, CA, USA). The procedure consisted of a reverse transcription of 7µg total RNA with a T7 oligo (dT) promoter primer, second-strand cDNA synthesis followed by an *in vitro* transcription to biotin-labeled cRNA. The cRNA concentration, purity, and size distribution were controlled again before the cRNA samples were hybridized on the arrays. Then, the labeled cRNA was fragmented and hybridized at 45°C for 16 hours on the Chicken Genome Arrays (Affymetrix, Santa Clara, CA, USA). The arrays contained 38,000 probe sets corresponding to more than 28,000 transcripts. The arrays were washed and stained using the GeneChip® Fluidics Station and scanned with a GeneChip® Scanner 3000 according to the manufacturer's protocol (Affymetrix Inc., Santa Clara, USA). The GeneChip® operating software (GCOS, Affymetrix, Santa Clara, CA, USA) controlled the scanner, the fluidics station, and the pre-normalization of the data.

Reverse transcription and real-time PCR

To verify the results from the microarray experiment, the total RNA was extracted from the remaining liver samples. RNA extraction, quantification, and quality control were performed equally for the microarray samples. The experiments for 17 samples per group were conducted in two steps: first, the reverse transcription was performed using the High-Capacity cDNA Reverse Transcription Kit (Applied Biosystems, Rotkreuz, Switzerland) and, second, the real-time PCR (qPCR) was conducted with adjusted amounts of cDNA. Primers were designed with Primer Express® 3.0 software (Applied Biosystems, Rotkreuz, Switzerland) on two adjoining exons in order to exclude a possible amplification due to genomic DNA contamination. *Glyceraldehyde-3-phosphate dehydrogenase* (*GAPDH*) and *lactate dehydrogenase A* (*LDHA*) were used as endogenous controls for adjusting the cDNA concentration. Those genes had a low standard deviation in the microarray experiment between the samples and the groups. Genes and the corresponding primer pairs are indicated in table 4.2. The 10-µl real-time PCR reaction contained 500nM of forward and reverse primers, 100 ng of sample cDNA, and 5 µl Fast SYBR Master Mix (2X) (Applied Biosystems, Rotkreuz, Switzerland). The PCR amplification was performed on the 7500 Fast Real-Time PCR System with SDS Software 1.3.1 (Applied Biosystems, Rotkreuz, Switzerland). The cycling program was 20 sec at 95°C and then 40 cycles of the following: denaturation for 3 sec at 95°C and annealing and extension at 60°C for 30 sec. The results were evaluated with the comparative Ct method ($2^{-\Delta\Delta C_t}$) with the SDS Software.

Table 4.2: Oligos for real-time PCR

Gene	Forward	Reverse
AKR1B1	TTAAAGAGATTGCAGCCAAGCA	TCACGTTTCTCTGGATGTGGAA
ALDH4A1	GCACATCTCCATGTACTCGTGAA	CAATAGGGCAGATAGAGCTGGAA
APOB	CAAAGCTTCCAGTGTTGGTTCA	TGGTTTTGGCCCAAGTGATATC
CDC2	CCATGGAAAGGAATTCAAAAATG	AGCTGCATCATCCCAATATAGTCTG
CPT1A	CGTCCCGAGAAGAGTTTTACAAG	TGCTGAACACGGCAAACTTT
CYP1A1	CGGTCAGGATTTGCAAGAGAC	CAGCTTCCAGGACCTGCAG
CYP1A4	TGGCAAGCGAACCACGA	GTGCCGGCTGCATTGAG
CYP4B1	AACTCAGCACACAGGCATTCAC	GCTGCCCAATGCAGTTCCT
CYP4V2	TCTGCTGGCCCCAGGAA	AGAAAACTCTTCTAGCCCTCATCCT

EME1	CACGTGTGCAGAGTTGCAAA	GATCCTGGCCGGGAGATAA
GAPDH	TTGGCATTGTGGAGGGTCTT	GGGCCATCCACCGTCTTC
GJA1	CCCAGGGCACTCCAATCAC	GTTTTCCCTTAACCCTCCAGAGA
KLF10	GGTCAGGCAAAAGGCGG	TGAGACCCATCACGCCG
LDHA	AACAAGATCAGCGTGGTTGGT	AAGGGTAAGTTCATCTGCCAAGTC
MRPS18A	GGGAAGCGGCTGGGTTT	ATTATGGTCGTGTTGCCTTCAGT
MT4	CAGTGGCAGCAGCTGCACT	GCTGCAACAACTGTGCCAAG
SGK2	AGCATAAAAAGTCCCATCACTCTTG	GAAGGTTATTGGCAAAGGAAGCT
SLC16A1	GCAGGGTTCAGGTTAAATGCA	CACCGTGGAGGAGCTCTACTTC
SLC27A4	TCCTGCCTTCCCCTCAATG	CGCATCCTCAACCTGACAGA
SMC2	GAGAAGGATGGCTAAGATCAAGGAC	CTGGAAGGAGAACTTAACAGAACTTAGTG
SOCS2	AGTTCAGGCTGGACTCCATCA	GGTGCACGACGCTGTTGA
SOD3	CAGCACGCCAGACGTTTG	ACATGGGCAAGGGCAATAAC

Sequence: 5' – 3'

High-throughput, real-time PCR

Some of the genes were tested with the high-throughput, real-time PCR after the evaluation of the assays with the real-time PCR on the 7500 Fast Real-Time PCR System (Applied Biosystems, Rotkreuz, Switzerland). The assays are indicated by an asterisk in table 4.4. As template cDNA, 16 samples from both control and MD20 and 15 samples from D20 of the real-time PCR experiment were used. First, a preamplification with the TaqMan® Pre-Amp Master Mix Kit (Applied Biosystems, Rotkreuz, Switzerland) was performed according to the protocol of Applied Biosystems with modification of Fluidigm (South San Francisco, CA, USA). The final reaction with a total volume of 5 µl contained 2.5 µl TaqMan® PreAmp Master Mix (Applied Biosystems, Rotkreuz, Switzerland), 0.05 µM of pooled primer mix, and 62 ng of cDNA. The pooled primers consisted of a mix of all 16 primers used in the later real-time PCR with a concentration of 0.2 µM for each primer. For the preamplification of the target fragments, the following cycling protocol was used: 10 min at 95°C and 14 cycles of 15 sec at 95°C and 4 min at 60°C. For the real-time PCR the final sample mix of 6 µl contained 3 µl of 2X TaqMan® Gene Expression Master Mix (Applied Biosystems, Rotkreuz, Switzerland), 0.3 µl 20X DNA Binding Dye Sample Loading Reagent (Fluidigm South San Francisco, CA, USA), 0.3 µl 20X EvaGreen™ (Hayward, CA, USA), 0.9 µl 1X TE Buffer, and 1.5 µl of a 1:5 dilution (1X TE buffer) of preamplificated cDNA. The Assay Mix with a volume of 6 µl contained 5 µM of forward and reverse primers, 3 µl of 2X Assay Loading Reagent (Fluidigm South San Francisco, CA, USA), and 1.5 µl of 1X TE buffer. For the real-time PCR, a 48.48 Dynamic Array (Fluidigm South San Francisco, CA, USA) was used with the priming station IFC Controller MX and the light cycler BioMark™ system, both from Fluidigm (South San Francisco, CA, USA). For the priming, 5 µl of the sample mix and 5 µl of the assay mix were loaded into the corresponding wells from the chip. A total of 47 samples and one negative control with water and 16 different primer assays with three replications were loaded, resulting in 2304 single real-time PCRs. After the priming, the chip was run according to the following cycling protocol: 2 min at 50°C, 10 min at 95°C, and 35 cycles starting with 15 sec at 95°C followed by 1 min. at 60°C. At the end of the cycling protocol, a dissociation stage was added to control for the specificity of the primers. Fluidigm BioMark Data Collection and Fluidigm Real-Time PCR analysis software programs were used to control the light cycler and to analyze the results. For the gene expression, the comparative Ct method ($2^{-\Delta\Delta Ct}$) was used as well. Genes *GAPDH* and *LDHA* were finally used as endogenous controls after evaluation with geNorm® (Ghent University Hospital Center for Medical Genetics, Ghent, UK). Other endogenous control genes tested were *albumin, dehydrogenase/reductase (SDR family) 7, lysosomal protein transmembrane 4β,* and *actin*.

Glucose, cholesterol and LDL-cholesterol measurement

The glucose, cholesterol, and LDL-cholesterol contents were measured in whole heparinized blood plasma. Due to the loss of two blood samples during slaughtering, the measurements were performed with 17 samples from the control, 20 from D20, and 18 from MD20. The samples were analyzed with the Cobas c 111 (Roche Diagnostics AG, Rotkreuz, Switzerland) with the corresponding kits from Roche for cholesterol (CHOL2), LDL-cholesterol (LDL_C), and glucose (GLUC2) (Roche Diagnostics AG, Rotkreuz, Switzerland). The results from the blood plasma were correlated to the normalized expression values from the microarray analysis of genes related to the measured substance.

Processing and statistical analysis of microarray data

The microarray images were transformed into intensity values for each probe set using MAS 5.0 software (Affymetrix, Santa Clara, CA, USA). All microarrays attained the acceptable Affymetrix quality control criteria. The processing of the intensity values was carried out using robust multi-array analysis (RMA), which was described by Irizarry *et al.* (2003) [175], with the GeneSpring GX 10 software (Agilent Technologies, Basel, Switzerland). Probe sets with a lower raw value than the 63.4th percentile, which corresponds to the raw value of 100, were not used for the gene expression analysis. The remaining genes were filtered on a volcano plot with a p-value below 0.05 and a fold change (fc) of at least 1.2. For the identification of the endogenous control genes for the real-time PCR, the initial gene list was filtered on the expression level of the raw value at the 99th percentile. Genes with the lowest deviation between all samples and the groups were selected. All of the other data was analyzed using ANOVA with R statistical software [176] to determine significance ($P < 0.05$).

Pathway and transcription factor analysis

The online program MetaCore™ (GeneGo, St. Joseph, MI, USA) was used to analyze the belonging of the genes to the different pathways and networks. The transcription factors were identified by the relation of significantly regulated genes to all the genes regulated by the corresponding transcription factors. The transcription factors of a high number of genes are known. For every gene, all the known transcription factors are available in public databases. When all the transcription factors are add up from every gene of one gene list, some transcription factors are over-represented. The probability is high that those transcription factors are activated under the experimental conditions.

For this calculation, the probability was calculated that a certain transcription factor is present in the final list and the false discovery rate (FDR) had to be below 0.05. Transcription factors with a lower expression than the 63.4 percentile in 11 of 22 samples were not used.

Results

Diet characterization and performance of broilers

In table 4.1, the results are given for the analysis of dry matter, crude ash, crude fat, crude protein, and gross energy of the broiler feed. All results were close to standardized diets (National Research Council, 1994) or within the range of the standard deviation. All broilers were in a good health and did not show any sign of intoxication. During the feeding trial, a total of three broilers died. Two broilers died in week three, one from the control and one from MD20. In the last week, one broiler died from MD20. None of the broilers from D20 died during the feeding trial. The mycotoxin analysis of the feed mixture from D20 and MD20 resulted in a slightly lower concentration of 17.7 mg DON/kg feed. For simplification, the groups are not renamed (D20/MD20). Zearalenone was detected on a very low level. The mycotoxin T-2 and HT-2 were below detection limit of 0.1 mg/kg feed.

Table 4.3: Growth parameters of broilers

	Control [A]		D20 [B]		MD20 [C]				
	Mean [g]	*s.e.*	Mean [g]	*s.e.*	Mean [g]	*s.e.*	A x B	A x C	B x C
BW at d 0	45.2		45.4		45.4				
BW at d 7	159.1		153.5		162.0				
BW at d 14	467.9	3.6	447.5	4.9	477.9	7.7	**0.016**	NS	**0.016**
BW at d 21	1023.0	13.9	964.1	13.7	1002.4	17.6	**0.023**	NS	NS
BW at d 28	1698.4	37.5	1619.2	26.1	1686.8	35.3	NS	NS	NS
BW at d 35	2395.3	42.4	2318.5	108.7	2368.8	45.1	NS	NS	NS
BWG week 1	113.9		108.1		116.6				
BWG week 2	308.8	3.6	294.0	4.9	315.9	7.7	NS	NS	NS
BWG week 3	555.2	12.8	516.6	9.0	524.5	12.9	**0.049**	NS	NS
BWG week 4	675.4	26.8	655.1	24.8	684.4	18.5	NS	NS	NS
BWG week 5	696.9	26.9	699.4	56.1	682.1	46.9	NS	NS	NS
FC week 1	142.0		126.7		139.3				
FC week 2	437.1	27.6	382.6	13.0	405.8	9.1	NS	NS	NS
FC week 3	844.6	43.8	715.3	12.0	743.4	17.3	**0.029**	NS	NS
FC week 4	1101.9	26.7	989.0	56.1	1076.8	14.1	NS	NS	NS
FC week 5	1230.9	29.5	1144.8	72.9	1164.9	11.7	NS	NS	NS
Total FC	3756.5	61.9	3358.3	98.1	3530.1	44.0	**0.014**	**0.025**	NS
Total FCR [g/g]	1.57	0.015	1.48	0.040	1.52	0.028	NS	NS	NS
Spleen weight [g]	2.19	0.097	2.13	0.103	1.87	0.118	NS	NS	**0.043**
Rel. spleen W. [g/kg]	0.94	0.045	0.959	0.045	0.795	0.039	NS	**0.041**	**0.019**
AME P1 [MJ/kg]/[%]	14.32	0.11	14.33	0.18	14.34	0.11	NS	NS	NS
AME P2 [MJ/kg]/[%]	13.86	0.13	14.26	0.06	13.66	0.18	**0.033**	NS	**0.02**
N Retention P1 [%]	63.56	1.21	63.21	0.89	63.72	1.17	NS	NS	NS
N Retention P2 [%]	56.86	2.04	59.59	3.02	54.9	2.12	NS	NS	NS

BW: Body weight, FC: Feed consumption, FCR: Feed conversion rate, Rel. spleen W: Relative spleen weight AME P1/2: Apparent metabolizable energy in MJ/kg dry matter and % relative to the control. P1/2: Period 1 or 2. N: 4 cages per group

A reduced body weight was observed at day 14 in D20 compared to the control and MD20 (table 4.3, fig. 4.1). After the third week, the body weight in D20 was still significantly reduced but only in comparison to the control. After the measurement at the end of the third week until the end of the feeding period, the body weight did not differ among the groups. The single animal weight before the slaughtering did not result in significantly altered body weight, although the birds in D20 had the lowest weight followed by MD20 and the control. A numerically increased standard error was observed in D20 at the end of the experiment.

Figure 4.1: Broiler growth

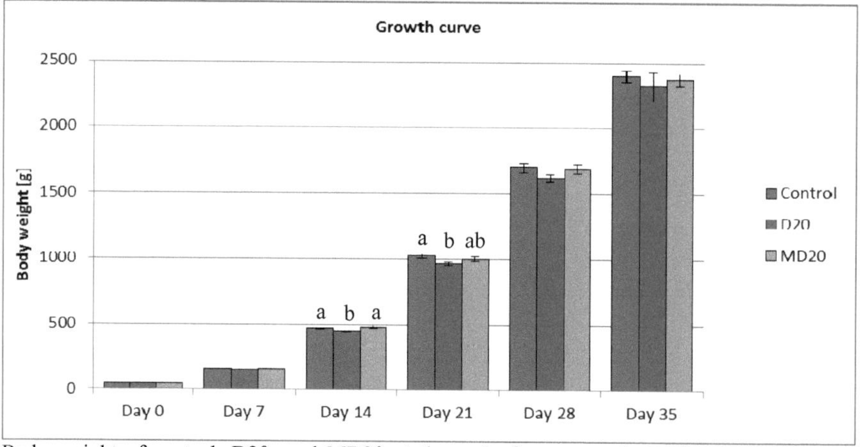

Body weight of control, D20, and MD20 at the end of every week. Different letters (a, b) are signficantly different ($P < 0.05$).

The weekly weight gain was signficantly reduced in group D20 in week 3. During the entire period, the weight gain of D20 was numerically lower compared to the control and MD20, with the exception of the last week. The contamination of the feed also led to an increased standard error of the weight gain, especially in D20.

Figure 4.2: Weight gains of broilers

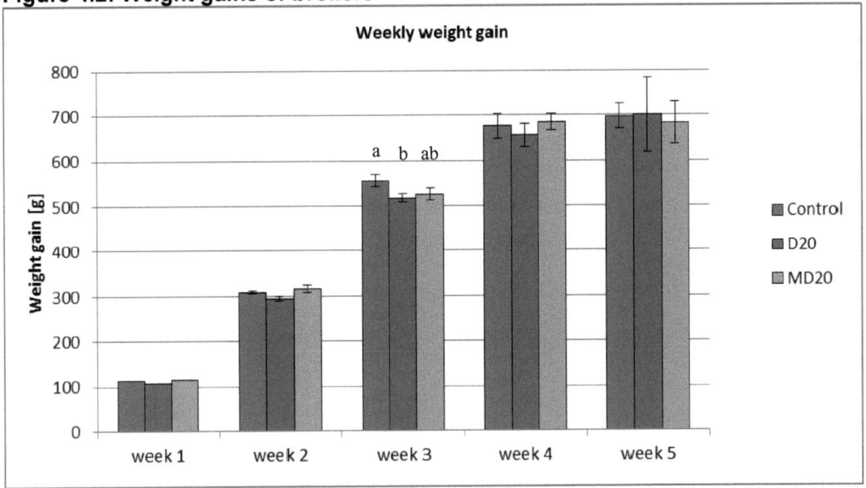

The weekly weight gain was altered in week 3 between the control and D20. Different letters (a, b) are signficantly different ($P < 0.05$).

The feed consumption was significantly lower in week 3 in D20 compared to the control. In total, the broilers consumed significantly less feed in D20 and MD20 over the feeding trial compared to the control. The feed conversion rate (FCR) did not differ significantly in any week or during the experiment, although the highest rate was in the control followed by MD20 and D20. The absolute spleen weight was significantly reduced in MD20 compared to D20, and the relative spleen weight was significantly reduced in MD20 compared to both the control and D20.

In the second excreta sampling period, the apparent metabolizable energy (AME) was significantly increased in D20 compared to the control and the MD20. The AME was significantly reduced in the second sampling period compared to first for the control and MD20, but it was similar for D20. The N retention did not vary among the groups.

Figure 4.3: Apparent metabolizable energy

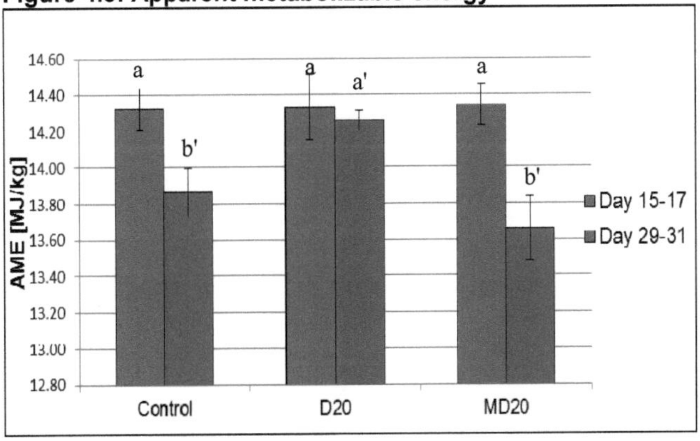

The apparent metabolizable energy (AME) was signficantly altered in D20 compared to the control and MD20 in the second sampling peiod. The AME was reduced in the second measurement period of the control and MD20, but was similar for D20. Different letters (first period: a, b, second period: a', b') are signficantly different ($P < 0.05$).

High-density oligonucleotide array hybridization

The parameters for the RNA quality and quantity, as well as the microarray quality parameters, were within the requirements. Elevated variances in the gene expression profile among the birds within D20 were observed and are presumably due to the use of commercial broilers.

Figure 4.4: Venn diagram of significantly altered probe sets

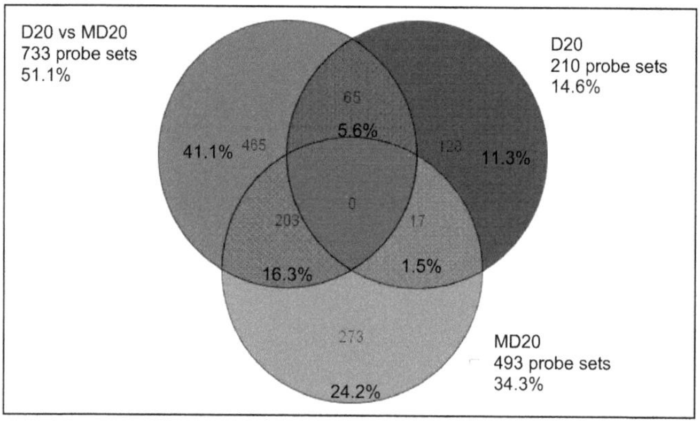

Outside of the Venn diagram the number of genes with a significantly altered expression is indicated. The percentage of differently expressed genes is based on all significantly regulated genes in all comparisons.

In comparison to the control, D20 had only 210 probe sets and 195 genes, which were significantly regulated. Some genes have several probe sets that encode for the same gene on a different position in the gene sequence. In MD20, 493 probe sets and 433 genes were significantly regulated. Only 17 genes were significantly regulated in D20 as well as in MD20. The highest number of significantly regulated genes was found between D20 and MD20 with 635 genes or 733 probe sets. Of those genes, 16.3% and 10.6% of the genes were significantly regulated in MD20 and D20, respectively. USP5 and BRWD1 were measured with several probe sets on the microarray. Of those two genes, different probe sets were significantly regulated in all comparisons. The significantly regulated genes in D20 were related to apoptosis, cell junction, cell cycle, cytoskeleton organization, endocytosis, ER to Golgi transport, immune system, lipid and fatty acid metabolism, signal transduction, transcription, translation, transport, ubiquitin proteolytic pathway, and the xenobiotic metabolism (table 9.1, appendix C).

In MD20, the genes with an altered expression were related to the actin cytoskeleton, apoptosis, cholesterol and lipid metabolism, cell cycle, cell junction, DNA repair, endocytosis, vesicular transport, immune system, insulin signaling, RNA processing, signaling, translation, and transport (table 9.2, appendix C).

Genes from the comparison of D20 to MD20 were related to apoptosis, cell cycle, cell junction, cholesterol, cytoskeleton organization, endocytosis, energy metabolism, ER to Golgi transport, fatty acid metabolism, immune system, insulin signaling, metabolism, RNA processing, signal transduction, transcription, translation, transport, ubiquitin proteolytic pathway, and xenobiotic metabolism (table 9.3, appendix C).

Real-time PCR

In total, 20 genes were selected for the verification with real-time PCR (table 4.4). Of these selected genes, 18 were significantly regulated in the microarray analysis. The additional genes *EME1* and *SLC27A4* were regulated in the previous experiment with up to 5 mg/kg DON. From the 18 genes of the microarray analysis, 16 genes could be confirmed with real-time PCR (88.9%).

Table 4.4: Real-time PCR results

	Liver											
	Microarray						qPCR					
	D20		MD20		D20 x MD20		D20		MD20		D20 x MD20	
	Fc	P	Fc	P	Fc	P	Fc	P	Fc	P	Fc	P
SOD3	1.841	0.0023		NS		NS	1.330	0.035	1.475	0.022	1.109	NS
*CYP1A4**	1.447	0.0041		NS	-1.735	0.0046	1.5641	0.0021	1.0206	NS	-1.523	0.0028
*CYP1A1**	1.516	0.0074		NS	-1.354	0.0114	1.9747	0.0013	1.5805	0.0382	-1.250	NS
*SMC2**	1.488	0.0022		NS	-1.664	4.10E-04	1.435	0.026	1.7103	NS	-1.192	NS
CDC2	1.381	0.0222		NS	-1.491	0.0044	1.500	0.0408	1.426	NS	-1.022	NS
ALDH4A1	1.207	0.005		NS		NS	-1.191	NS	-1.047	NS	1.138	NS
*CYP4V2**	-1.298	0.0159		NS	1.506	0.0278	-1.301	NS	1.3574	NS	1.766	0.0055
*SLC16A1**	-1.464	0.0476		NS	1.363	0.0297	-1.22	0.0333	1.0442	NS	1.273	0.0334
*MT4**	-2.186	0.0059		NS		NS	-2.139	0.013	1.4001	NS	2.995	0.0002
*GJA1**		NS	-1.823	0.0477	-1.806	0.038	-1.178	NS	-1.517	0.0056	1.2878	NS
SOCS2		NS	-1.949	0.009	-1.597	0.021	-1.438	0.0255	-1.722	0.0003	1.197	NS
KLF10		NS	-1.553	0.002	-1.538	0.003	1.020	NS	-1.116	NS	-1.138	NS
CPT1A		NS		NS	3.020	0.029	1.475	NS	2.547	0.007	1.726	NS
CYP4B1		NS		NS	2.175	0.022	-1.049	NS	1.882	0.029	1.800	0.04
*APOB**		NS		NS	2.162	0.013	1.021	NS	1.236	0.0306	1.223	0.0453
*MRPS18A**		NS		NS	-1.213	8.36E-04	1.103	NS	-1.235	0.0257	-1.363	0.002
SGK2		NS		NS	-1.239	0.0433	1.125	NS	1.480	0.020	1.315	NS
AKR1B1		NS		NS	-2.223	0.0131	-1.148	NS	-1.692	0.007	-1.474	0.043
*SLC27A4**		NS		NS		NS	-1.382	NS	-1.352	NS	-1.022	NS
EME1		NS		NS		NS	1.176	NS	1.089	NS	-1.080	NS

*: Gene expression measured with BioMark™ system; SLC16A1 and MT4 expression change in D20 microarray with different normalization method

N = 13 per group

NS: not significant, $P < 0.05$

Transcription factor analysis

With the three different gene lists from D20, MD20, and the comparison of D20 vs. MD20, the transcription factors with the highest number of genes they are influencing were detected (table 4.5). In the "r" column, the number of genes with a significant expression change and the corresponding transcription factor from the first column are indicated. Column "R" indicates the number of genes influenced by the corresponding transcription factor. The column "Expected" indicates how many genes would be influenced by the corresponding transcription factor by chance in relation to the size of the gene list. The ratio is defined as the number of genes influenced by a transcription factor (r) divided by the expected number of influenced genes (Expected). The p-value shows the probability to have the given or a higher value of the column "r". The z-score ranks the relation of genes from the gene list with significantly altered expression (column "r") to the total number of genes, which are influenced by a transcription factor (column "R").

Table 4.5: Transcription factor analysis

Transcription Factor	Function	r	R	Expected	Ratio	p-value	z-score
Transcription factors of D20							
E2F2	Cell cycle	4	64	0.4152	9.633	7.975E-04	5.589
E2F6	Cell cycle	4	71	0.4606	8.684	1.178E-03	5.24
E2F4	Proliferation inhibit.	9	276	1.791	5.026	8.181E-05	5.44
E2F1	Cell cycle	11	522	3.387	3.248	6.122E-04	4.202
Transcription factors of MD20							
SREBF2*	Cholesterol homeostasis	6	55	0.734	8.174	8.981E-05	6.196
SREBF1	Sterol biosynthesis	13	200	2.669	4.87	3.111E-06	6.395
E2F1	Cell cycle	20	522	6.967	2.871	2.639E-05	5.032
C/EBPbeta*	Immune response	19	534	7.127	2.666	1.116E-04	4.534
EGR1	Differentiation	15	436	5.819	2.578	8.229E-04	3.871
c-Jun	Signalling	15	463	6.179	2.427	1.494E-03	3.611
CREB1	Immune response	25	988	13.19	1.896	1.725E-03	3.353
ESR1	Hormone receptor	27	1083	14.45	1.868	1.406E-03	3.409
Transcription factors of D20 vs MD20							
E2F6	Cell cycle	9	71	1.487	6.052	1.690E-05	6.236
E2F2	Cell cycle	7	64	1.341	5.222	3.747E-04	4.947
NFYB	Antigen presentation	7	73	1.529	4.578	8.365E-04	4.479
CLOCK	Circadian rhythm	7	75	1.571	4.456	9.833E-04	4.385
HNF3-beta	Diabetes related	18	226	4.734	3.802	1.458E-06	6.195
FOXO1	Gluconeogenesis	8	103	2.157	3.708	1.460E-03	4.03
E2F4	Proliferation inhibit.	20	276	5.781	3.46	1.684E-06	6.015
AHR	Xenobiotic metabol.	20	302	6.326	3.162	6.585E-06	5.533
C/EBPbeta*	Immune response	27	534	11.19	2.414	2.561E-05	4.839
HNF1-alpha	Glucose homeostasis	21	419	8.776	2.393	2.260E-04	4.211
E2F1*	Cell cycle	26	522	10.93	2.378	4.638E-05	4.661
ESR1 (nuclear)	Estrogen receptor sig.	43	1083	22.68	1.896	4.664E-05	4.423
c-Myc	Cell cycle arrest	63	1681	35.21	1.789	4.639E-06	4.929
CREB1	Immune response	35	988	20.69	1.691	1.795E-03	3.253
HNF4-alpha*	Xenobiotic metabol.	77	2339	48.99	1.572	3.926E-05	4.282
SP1	Apoptosis	73	2341	49.04	1.489	3.379E-04	3.663

*: Gene with significantly altered expression
All transcription factors had an expression value of at least the 63.4th percentile in 11 of 22 samples.
r: Number of genes sig. regulated in D20 by corresponding transcription factor

N D20: 140 significantly regulated gene-based objects from D20
N MD20: 288 significantly regulated gene-based objects from MD20
N D20 vs. MD20: 452 significantly regulated gene-based objects from comparison D20 vs. MD20
R: Number of genes regulated by corresponding transcription factor
N: 21579 gene-based objects in the complete database
Expected: Mean value for hypergeometric distribution (n*R/N)
Ratio: Connectivity ratio (r/Expected)
Z-score: The Z-score ranks the sub-networks of the analyzed network algorithm with regard to their saturation with genes from the experiment (r-Expected)/sqrt(variance).
False discovery rate: 0.05

Glucose, cholesterol, and LDL-cholesterol concentration in heparinized blood plasma

The glucose concentration in heparinized blood plasma was not significantly altered. The highest concentration was observed in D20 followed by the control and MD20. The correlation between glucose concentration and the different insulin signaling related genes and was the highest for *PPARA* (-0.68), *PPARGC1A* (-0.67), *IGFBP1* (-0.63), *INSR* (-0.62), *CEBPB* (0.62), and *GRB2* (0.60).

Table 4.6: Glucose, cholesterol, and LDL-cholesterol in heparinized blood

Substance	Control [A]		D20 [B]		MD20 [C]				
[mmol/L]	Average	*s.e.*	Average	*s.e.*	Average	*s.e.*	AxB	AxC	BxC
Glucose	14.09	0.31	14.19	0.19	13.79	0.33	NS	NS	NS
Cholesterol	3.13	0.11	3.23	0.08	3.08	0.06	NS	NS	NS
LDL-cholesterol	0.558	0.035	0.657	0.056	0.682	0.040	NS	0.026	NS
LDL-C/cholesterol	0.180	0.008	0.200	0.013	0.220	0.012	NS	0.013	NS

s.e.: Standard error
LDL-C/cholesterol: Proportion of LDL-cholesterol on total cholesterol
The measurement unit was mmol/L for all results.
N: Control: 17, D20: 20, MD20: 18

The cholesterol concentration in the plasma was not significantly altered. Although a numerical decrease in MD20 (-1.6%) and a numerical increase in D20 (+3.4%) was observed. With the removal of one outlier sample from both D20 and MD20, the significance of decreased cholesterol content would have been achieved between D20 and MD20 (-4.6%). The comparison of samples used in the microarray analysis between D20 and MD20 resulted in a p-value of 0.071. The correlation between the expression of the rate limiting enzyme in cholesterol biosynthesis of *HMGCR* and the cholesterol concentration in blood plasma was only 0.18. The highest correlation was observed for *MVD* (0.48), *LSS* (0.47), *DHCR24* (0.44), and *SREBF2* (0.40). *GOPC* was negatively correlated with the cholesterol concentration (-0.59).
The LDL-cholesterol concentration was significantly increased between the control and MD20 by 22.2%, as well as the proportion of LDL-cholesterol on the total cholesterol by 22.0%. All the other comparisons were not significantly altered. Numerically, the lowest LDL-cholesterol concentration was in the control followed by D20 and MD20. The gene responsible for LDL-cholesterol uptake *(LDLR)* had a significantly decreased expression of -1.70 in MD20 compared to the control as well

as between D20 and MD20 (-1.70). The correlation of *LDLR* expression and LDL-C concentration was -0.17, for *APOB* -0.34, and the highest correlation was observed for *MVD* with 0.39.

Figure 4.5: Glucose, cholesterol, and LDL-cholesterol concentration

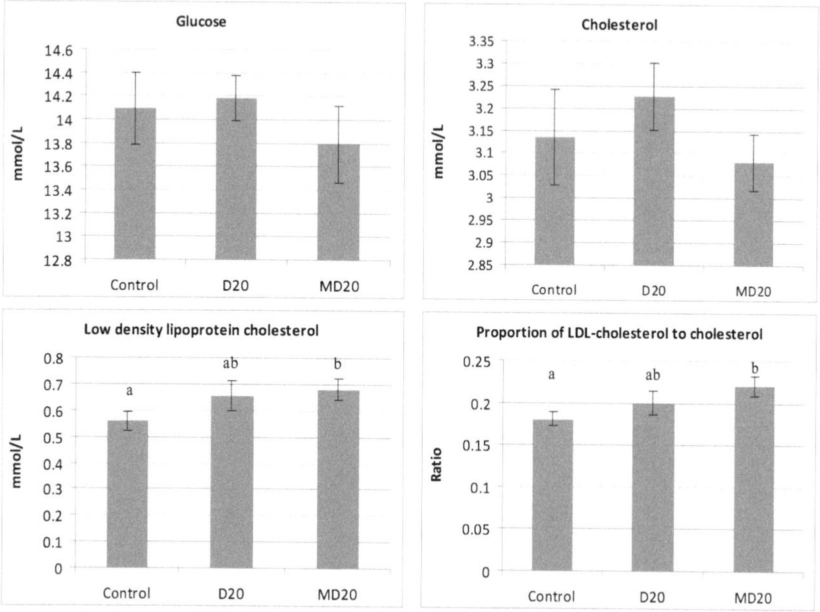

Glucose, cholesterol, and low-density lipoprotein cholesterol measurement in heparinized blood
Significant differences (P < 0.05) are indicated by different letters.
N: control and D20: 17, MD20: 18

Discussion

Diet characterization and performance of broilers

After two weeks, the average body weight of D20 was significantly reduced compared to the control and the MD20. A significant reduction was also observed after the third week in D20 compared to the control, but no longer in comparison to MD20. After weeks four and five, no significant difference was observed, although the average body weight in D20 was numerically lower compared to the control and MD20. This indicates that the D20 birds and their metabolism adapted to the DON concentration and partially recovered from the detrimental effects of DON. Especially in week 5, the weekly weight gain was numerically higher compared to the other groups. The standard error of weight gain and body weight in D20 increased compared to the other groups and the weeks before, showing that the adaption might be influenced by the genetic background of the birds. The numerically lower body weight of the MD20 birds compared to the control indicates that the dose reduction of DON, due to the binding to the mycotoxin absorber, was not high enough for a complete immobilization of DON in the intestine. On the other hand, in the first three weeks, the body weights were higher in MD20 compared to the control group.

Awad *et al.* (2004) reported that broilers fed 10 mg DON/kg feed had a higher body weight than the control in the first two weeks of the experiment [177]. Swamy *et al.* (2004) observed a linear decreased feed consumption and body weight gains as the DON concentration in the feed increased up to 9.7 mg DON/kg feed during the grower phase (days 21-42) [171]. Our results were similar in D20, but the decreased feed consumption and body weight were observed from day 14 to day 21, which might be due to the faster growth of our birds; their weight gain up to day 14 was similar to the weight gain until day 21 of Swamy *et al.* (2004). They also mentioned that an adaption to the mycotoxin could occur over time [171]. This adaption might be in relation to the xenobiotic metabolism-related genes, which will be discussed in relation to the D20. The feed consumption was significantly reduced in D20 in the third week and over the feeding experiment. Also, the MD20 birds showed a reduced total feed consumption over the entire feeding experiment similar to D20. This and the numerically reduced body weight indicate that the supplementation with the glucomannan polymer should be increased to immobilize DON with a concentration of 17.7 mg DON/kg feed. According to our results, the polymeric glucomannan myctoxin absorber has a diminishing influence on the amount of DON absorbed by the animal.

The recovery of the birds from D20 seems to be directly related to the significantly increased apparent metabolizable energy (AME) in the beginning of the fifth and last weeks. The AME measurements in the second period of the control and MD20 were significantly reduced from the first period. In contrast, the AME of D20 was similar to the measurements of the first period. It seems that the development of the birds from D20 was delayed compared to the other groups, which would also explain the significantly altered expression of genes related to the cell cycle, which is discussed below. The nitrogen retention was only numerically increased in relation to the control and MD20.

The very low number of genes that are regulated in the two groups MD20 and D20 (1.5%, fig. 4.4) indicate that the effect of 17.7 mg DON seems to be completely different when the mycotoxin absorber is added. Also, the high amount of significantly altered genes in the comparison between D20 and MD20 indicate a completely different function scheme. Of all probe sets regulated between D20 and MD20, 37.6% are also regulated between control and the D20 or MD20, showing a two-sided confirmation. This indicates that the D20 and MD20 are differently affected by the DON contamination, due to the mycotoxin absorber. The low number of significantly altered genes in D20 seems be related to the increased standard error of the body weight in this group. Interestingly, the bird with the highest BW was from the control group followed by two birds from D20. The adaption of the birds of D20 to DON might be based on altered microbial composition in the intestine, on the genome of the birds, or on the stage of development. Based on our results, the alternate feeding of contaminated and uncontaminated feed with a weekly change would probably show higher performance differences among the groups. This experimental design would also represent to a high degree the situation in commercial broiler production. Depending on the cereal batch, the DON concentration can vary strongly, which would avoid an adaption of the birds to DON in commercial broiler production.

The reason for the reduced relative spleen weight in the MD20 compared to the control and D20 remains unclear. Ozkan *et al.* (2007) reported a reduced relative spleen weight in broilers supplemented with organic selenium under cold conditions [178]. Puvadolpirod and Thaxton (2000) showed a reduced spleen weight and body weight when boilers were challenged with the physiological stressor adrenocorticotropin (ACTH) for seven days [179]. The two studies indicate that the reduced spleen weight might be a result of physiological stressors. In relation to the spleen, Pestka *et al.* (1990) reported a significantly increased percentage of T cells in murine spleens due to the feeding of 25mg DON / kg feed [180]. Swamy *et al.* (2004) did not find any changes in organ weight when broilers were fed with up to 11.3 mg DON/kg feed. They concluded that organ weight is not a good indicator for toxicity [171]. It remains unclear what caused the reduced spleen weight in MD20.

Xenobiotic metabolism-related genes

Cytochrome P450, 1A1 (fc: +1.516, *CYP1A1,* D20, D20 vs. MD20) was upregulated in D20 in the microarray and the qPCR experiments and to a lesser extend in MD20 (fc: +1.58) only in the qPCR experiment. CYP1A1 is important in the metabolism of cyclic aromatic hydrocarbons [181]. In addition, it remains unclear how and to what extent CYP1A1 is involved in the detoxification of DON. Gouze *et al.* (2006) reported no changes in the liver of mice fed with pure DON in the protein expression of CYP1A1/2 family in contrast to the CYP2 family, which had an increased enzyme expression in the group with 0.2 and a 1 mg DON/kg diet, but not in the group with 5 mg/kg DON [170]. The difference to our result could be that they did not differentiate between CYP1A1 and A2.

AHR was detected in the transcription factor analysis (table 4.5, D20 vs. MD20) and mediates most of the toxicological effects of polycyclic aromatic hydrocarbons (PAH), possibly also by DON, which has an aromatic structure. AHR transcription factor might have influenced the expression change of *CYP1A1*, *CDC2* (cell cycle), and *CYP1A4*. AHR nuclear translocator (*ARNT*, fc: -1.11, D20) activates AHR by ligand binding, mediating transcription activation of *CYP1A1* [182].

The expression alteration of *CYP1A4* (fc: 1.56, D20, D20 vs. MD20) could be confirmed with qPCR. CYP1A4 is unique for chickens and it is a homologue to the human CYP1A1 and, like all CYP1A enzymes, it is transcriptionally induced [183]. PAPS synthase 2 (*PAPSS2*, fc: 1.21, D20) is responsible for the biosynthesis of PAPS, which is a sulphonate donor molecule. PAPS plays a crucial role in sulphonation of drugs and xenobiotics. *PAPSS2* transcription seems to be under the control of SP1 (table 4.5) [184]. *CYP4B1* (fc: 2.49, D20 vs. MD20) was significantly upregulated in MD20 compared to control and D20 in the qPCR experiment. CYP4B1 is involved in the synthesis of 12-hydroxy-5,8,11,14-eicosatetraenoic acid (12-HETE) and 12-hydroxy-5,8,14-eicosatrienoic acid (12-HETrE), which possess potent inflammatory properties [185].

In summary, *CYP1A1* was increased in D20 and to a lesser extent in MD20, but only in the qPCR experiment, and *CYP1A4* and *PAPSS2* were only increased in D20, but not in MD20. This would indicate that the mycotoxin absorber reduced the amount of DON reaching the liver. On the other side, the increased expression of *CYP4B1* indicates another additional mode of action with a possibly lower DON concentration or that the mycotoxin absorber itself influenced the gene expression. The mycotoxin absorber should bind mycotoxins in general; to what extent it also binds other substances has not yet been reported. Detoxification of DON has been mainly related to the bacterial degradation from DON to the de-epoxy-DON, the low absorption into plasma and tissue, and the rapid clearance [186]. The epoxide hydrolase enzymes (EPHX1/2) are phase I detoxification enzymes of xenobiotics. DON has a 12,13-epoxide group, but little is known about the effect of EPHX enzymes in its detoxification. It has been shown in the spleens of mice that the expression of *EPHX1* decreased after a single DON injection [187]. In our experiment, the *EPHX* genes were not significantly altered with exception for the comparison between the control and both DON-treated groups, in which *EPHX1* had a significantly reduced expression (fc: -1.14). The raw values of the *EPHX* genes were above the 93th percentile and *LOC421447*, which is similar to *EPHX1,* had a raw value above the 99th percentile. Therefore, the high expression of the epoxide hydrolase enzyme could be a reason for the insensitivity of chickens towards DON compared to other species. The comparison of the *EPHX* enzyme expression between the previous experiment with up to 5 mg/kg DON (23 days) and the current experiment did not show significant differences. Therefore, the adaption of the broilers in relation to the body weight from day 21 to day 35 cannot be explained by the *EPHX* genes in D20.

Oxidative stress-related genes

The superoxide dismutase 3 (*SOD3*, fc: 1.84, D20) and catalase (*CAT*, fc: -1.31, D20) are antioxidant enzymes. The two glutathione transferases (*GSTA1*, fc: -1.30, D20 vs. MD20, *GSTZ1*, fc: -1.24, D20 vs. MD20) play an important role in detoxification by catalyzing the conjugation of many compounds with glutathione. Gouze *et al.* (2006) showed that glutathione transferases might use DON as a substrate and target the 12,13-epoxide group [170]. In Caco-2 cells, it has been shown that oxidative stress is increased due to the administration of DON by increased amounts of malondialdehyde, which is biomarker of lipid peroxidation [78]. Furthermore, Bodea *et al.* (2009) showed that superoxide anion concentration significantly increased six hours after the treatment with DON and returned to control level after 24 hours in HepG2 cells. SOD and CAT activity also increased in the same experiment to a maximal specific activity after 24 hours. The glutathione concentration decreased over 24 hours, as well as the glutathione transferase and peroxidase activities [188]. The increased expression in our experiment of *SOD3* coincides with the results of Bodea *et al.* (2009). The *CAT* expression is significantly reduced in contrast to the activity increase of Bodea *et al.* (2009). Therefore, no change was observed in our study for the glutathione peroxidase, which also detoxifies hydrogen peroxide. The expression decreases of *GSTA1/Z1* in MD20 compared to D20 indicate that the DON induced oxidative stress seems to be dose

dependent. The increase of the superoxide concentration and its normalization shows the short-term effects of DON. The discrepancy to the study of Bodea *et al.* (2009) might be due to the *ad libitum* feeding of the birds resulting in the oscillation of the DON concentration in the animal in contrast to the sole DON administration in the cell culture. Interestingly, oxidative stress and apoptosis related genes were found in all three comparisons (tables 9.1-9.3, appendix C).
The aldo-keto reductase family 1, member B1 (*AKR1B1*, fc: -1.69, MD20, D20 vs. MD20) detoxifies reactive lipid aldehydes, which are formed in a greater extend upon oxidation of polyunsaturated fatty acids (PUFA) [69]. *AKR1B1* expression (fc: -3.53) and the enzyme activity (-18.5%) were significantly reduced in the liver of broilers upon supplementation of the feed with selenized yeast. The subsequent increase of PUFA in the liver showed reduced damage to PUFA, due to decreased oxidative stress. *AKR1B1* therefore seems to be a good marker for the amount of oxidatively damaged PUFA (Microarray analysis of different gene expression in low and adequate Se-fed poultry. Page 31). In addition, *AKR1B1* showed a reduced expression in the experiment with 2.5 and 5 mg/kg DON (fc:-2.08, fc: -2.50) (*Fusarium* mycotoxin-contaminated wheat containing deoxynivalenol alters the gene expression in the liver and the jejunum of broilers. Page 35). The increased expression of antioxidant and related proteins might have led to a reduced oxidative stress and the decreased expression of *AKR1B1*. Another explanation might be that the mycotoxin absorber also immobilized other substances, which can induce oxidative stress in liver cells. The expression of *AKR1B1* seems to be highly regulated in a diet-dependent manner, due to its strong regulation in all three experiments.

Cell cycle and proliferation

E2F transcription factors play an important role in cell cycle progression, DNA damage checkpoint and repair pathways, chromatin assembly and condensation, chromosome segregation, and mitotic spindle checkpoint [189]. At least seven E2F transcription factors exist. E2F1 (fc: -1.37, D20 vs. MD20) to E1F3 are mainly transcription activators. E2F4 and E2F5 (fc: 1.25, D20 vs. MD20) are related to transcription repression and bind mostly RB2 (unchanged) and RBL1 (fc: -1.27, D20 vs. MD20). Dependent on the cell cycle E2F genes are differently expressed, like *E2F1* and *E2F2* at the G1/S boundary. E2F4 and E2F5 are bound to the nucleus by RBL1 and RB2 at the G0 and early G1 phase [190]. Beside the altered expression of *E2F1*, *E2F5* in the comparison between D20 and MD20, E2F1, E2F2, E2F4, and E2F6 were the only transcription factors detected in the transcription factor analysis of group D20. This shows that the cell cycle might be regulated in D20. In mouse embryos, it has been shown that cell division cycle 2 (G1 to S and G2 to M) (*CDC2* (*CDK1*), fc: +1.50, D20) is necessary to undergo organogenesis. Santamaria *et al.* (2007) showed in mice, which were lacking the CDK's 2 to 6, that the cell cycle was not interrupted in embryonic fibroblasts. They also showed that CDK1 can bind to all cyclins resulting in the phosphorylation of retinoblastoma protein and the expression of genes, which are regulated by the E2F transcription factors (E2F1/2/4/6, table 4.5) [191]. CDK1 controls the G2 to M transition and the initiation of mitosis. In CDK2 knockout mice, CDK1 can functionally substitute for CDK2 and the cells progress from G1 to S phase. Furthermore, p21 seems to inhibit CDK1-cyclin complexes [192]. The

G2/M transition depends on the activity of CDC2. It has been suggested that genotoxin-induced G2 arrest might be associated with the inhibition of *CDC2* expression [193]. Diesing *et al.* (2011) reported in intestinal epithelial cells (IPEC-1 and IPEC-J2), challenged with up to 2μg/ml DON, a significant increase of cells cycle arrest in phase G_2/M after 72h. On the other side, the portion of cells in the G_0/G_1 phase was significantly decreased [168]. In human intestinal epithelial cells, *p21* expression was increased after the DON addition as well as the amount of cells in the G_2/M phase. In parallel the amounts of cells in the G_0/G_1 were decreased after 48h and the addition of up to 1μg/ml DON [169]. In conclusion, *p21* expression is significantly increased in cell cultures leading to the cell cycle arrest in G_2/M phase probably over the CDK1-cyclin complex inhibition. The upregulation of *CDK1* (*CDC2*) seems to restore normal cell cycle progression in our experiment. The cell cycle arrest after stress might allow the cell to re-establish homeostasis [169].

As mentioned in the section "Xenobiotic metabolism related genes," AHR (table 4.5, D20 vs. MD20) regulates the expression of *CYP1A1* and *A4* beside the indirect regulation of *CDC2*. In AHR null mouse embryonic fibroblasts, it has been shown that cell accumulate in the G_2/M phase, with a concomitant decreased proportion of the cells was in G_0/G_1. In parallel, the expression of CDC2 was decreased [182]. This indicates that AHR seems to be the linkage factor for the upregulation of xenobiotic metabolism-related genes and the recovery from the cell cycle arrest of the G_2/M phase. The upregulation of *CDC2* might counteract the cell cycle arrest, which is caused by DON. Yoon *et al.* (2010) showed that EIF2AK2 (unchanged) was activated by cyclo-hexamide and led to the ubiquitination of CDC2 and subsequently to its degradation. The expression level of *CDC2* was thereby not changed. They also showed that G_2 arrest over EIF2AK2 activation and CDC2 degradation was restored by the overexpression of *CDC2*. Further EIF2AK2 knockdown cells (HEK293) grew faster [193]. Interestingly, EIF2AK2 is activated by DON in cloned macrophages [194]. Several genes were found in relation to the G_2/M cell cycle progression in D20, which indicates that the alterations observed mainly in cell cultures might also occur in the living animal. The alterations of the discussed genes might also contribute to the recovery of the birds from the reduced body weight at the end of weeks 2 and 3.

Thymidine kinase 1 (*TK1*, fc: +1.32, D20) performs the transfer of the terminal phosphate of ATP to the 5' hydroxyl group of thymidine to form dTMP in cells for DNA synthesis. Its expression is cell-cycle regulated to coordinate with DNA replication and it is upregulated during G1/S transition on the level of transcription and translation. TK1 is phosphorylated by CDC2 during G_2/M phase, leading to a decrease of its catalytic capability and a negative modulation when DNA replication is complete [195]. Several other genes are phosphorylated by CDC2 like ribonucleotide reductase M2 polypeptide (*RRM2*, fc 1.56, D20, D20 vs. MD20), stathmin1 (*STMN1*, fc: 1.30, D20), topoisomerase II A and B (*TOP2A*, fc: 1.26, D20) and (*TOP2B*, fc: -1.29, D20), microtubule-associated protein 7 (MAP7, fc: -1.22, D20, D20 vs. MD20), TPX2 (fc: 1.48, D20, D20 vs. MD20), and cortactin (CTTN, fc: -1.21, D20) [196]. *Baculoviral IAP repeat-containing 5* (*BIRC5*, fc: 1.29, D20) expression seems to be regulated by CDC2. BIRC5 is an inhibitor of apoptosis [197]. It has been shown that casein kinase II beta subunit (*CSNK2B*, fc: -1.54, D20, D20 vs. MD20), which is part of the tetrameric casein kinase II complex, is phosphorylated by CDC2 in mitotic cells. Besides the expression change the relations of the raw values of CSNK2B to CSNK2A1 or A2 were also significantly reduced. CSNK2B is related to cell proliferation, differentiation, and survival. It also binds and phosphorylates the topoisomerase II [198], which had a significantly altered expression. Overexpression of *CSNK2B* leads to attenuated proliferation resulting from cellular G_2/M arrest [199]. Beside the upregulation of *CDC2,* the downregulation of *CSNK2B* might have contributed to the possible recovery from the G_2/M cell cycle arrest. According to Lin *et al.* (2009), phosphorylated STMN1 by CDC2 promotes microtubule polymerization and the phosphorylation is tightly linked to the progression of the cell cycle and cell proliferation [200]. The microtubule-associated, homolog TPX2 (fc: 1.48, D20, D20 vs. MD20) functions in spindle formation in the assembly of microtubules around chromatin and its activity is restricted to the M phase. TPX2 is inhibited by karyopherin alpha 2 (KPNA2, fc: 1.35, D20, D20 vs. MD20) by its binding. For the spindle assembly, TPX2 is released by KPNA2 [201]. PDZ binding kinase (*PBK*, fc: 1.41, D20, D20 vs. MD20) is phosphorylated by CDC2 and is a cell cycle-regulated kinase, which is only active at mitosis [202]. It has been shown that inhibiting the activity of RRM2 (fc: 1.56, D20, D20 vs. MD20) also inhibits cell growth. RRM2 is the rate-limiting enzyme in the conversion of ribonucleotides to deoxyribnucleotides. The expression levels of the *eukaryotic translation initiation factor 3 subunit A* (*EIF3A*, fc: 1.23, D20) and *RRM2* oscillate during cell cycle and peak in S phase and are decreased in G1 phase. In cell line H1299 RRM2, protein synthesis correlated with EIF3A levels and the downregulation of the *EIF3A* expression also led to decreased synthesis of RRM2 and DNA. Furthermore, in H1299 cells, the downregulation of EIF3A led to decreased [^{35}S] methionine incorporation [203]. The gene *RRM2* also has an E2F4 binding site and its transcription is influenced by E2F4 [204]. EIF3A and RRM2 seem to have an important influence on the increase of global protein synthesis and they seem to antagonize the inhibitory effect of DON on the translation.

Metallothionein 4 (*MT4*, fc: -2.139, D20) was significantly regulated in subset of samples between control and D20 and could be confirmed with qPCR. The strong decrease of the expression indicates that *MT4* expression regulation might be directly regulated by a DON-influenced transcription factor. MT4 from chicken genome is similar to human MT2A, meanwhile MT1 and MT2 are not available on the microarray and on Entrez database for the chicken genome. Metallothioneins bind primarily zinc and secondarily copper, but they have a higher affinity to toxic metals like Ag, Hg, Cu, and Cd when they are present. They are, in general, related to detoxification of heavy metals, homeostatic regulation of essential metals, and protection of tissues against oxidative injury [205]. The effect of zinc and copper on the gene expression of *MT4* seems to be marginal, although zinc and cadmium are important inducers of *MT4* transcription [206]. In addition, the copper transporter *SLC31A1* (fc: -1.28, D20) was slightly downregulated. Lim *et al.* (2009) [207] reported a slower cell growth in MT2A silenced MCF-7 breast cancer cells, an increased number of cells in the G1-phase, and a cell cycle arrest. They also showed that MT2A can mediate G1-growth arrest via ATM (fc: -1.16, D20)/CHK2/CDC25A pathway.

The structural maintenance of chromosomes 2 (*SMC2*, 1.44, D20) and 4 (SMC4: fc: -1.23, MD20) are subunits of condensin I and II. The condensin complex has a crucial role in the formation of stable mitotic chromosomes [208]. The MCM complex consists of the subunits MCM2-7 (*MCM2*, -1.63, MD20, D20 vs. MD20), (*MCM5*, fc: -1.57, D20 vs. MD20), (*MCM6*, fc: -1.63, MD20, D20 vs. MD20), and the MCM complex is used for the initiation of DNA replication and is loaded on the pre-replicative chromatin [209]. The downregulation of the *MCM* genes and the *SMC4* in MD20 compared to control and D20 indicates that DNA replication and cell division might also be affected.

Monolayer permeability

Cells are connected by intercellular complexes like tight junction, gap junctions, and adherens junctions [104]. In contrast to the previous results (*Fusarium* mycotoxin-contaminated wheat containing deoxynivalenol alters the gene expression in the liver and the jejunum of broilers. Page 37) from the experiment with up to 5mg/kg DON, the expression of the genes *TJP1, JAM2,* and *CLDN3* were not altered. Therefore, *the gap junction protein alpha 1* (*GJA1*, fc: -1.52, MD20, D20 vs. MD20) and *gamma 1* (*GJC1*, fc: -1.92, MD20) had a significantly altered expression. They connect adjacent cells to allow rapid exchange of nutrients, metabolites, ions, and small molecules of up to 1000 Da [210]. Therefore, the size of the gap junction would allow the passage of DON [211]. Claudin 2 (*CLDN2*, fc: -2.09, MD20, D20 vs. MD20) is part of the tight junctions and facilitate paracellular conductance [212]. The immunoglobulin superfamily 5 (*IGSF5*, (*JAM4*), fc: -1.22, D20 vs. MD20) is a cell adhesion molecule that interacts with tight junction protein and seems to be implicated in paracellular permeability [213]. De Walle *et al.* (2010) showed a decreased TEER and claudin 4 protein expression in Caco-2 cells as an *in vitro* model for the human intestinal barrier. They reported that the transepithelial electrical resistance (TEER) decreased to 21% after 24 hours of DON incubation at a concentration of 5µg/ml. It was mentioned that TEER reflects the functional tight junctions made of transmembrane proteins [104]. The decrease of the TEER shows the acute effect of DON on *in vitro* cell cultures. The extent to which the chronic administration of DON leads to same effects has not been reported. The significant expression change of *TJP1, JAM2,* and *CLDN3* from the first experiment with up to 5 mg/kg DON and the current experiment with the genes *GJA1*, *GJC1*, *CLDN2,* and *IGSF5* indicates that similar effects might also occur in the living animal.

Genes involved in hormone turnover and signaling

In our experiment, the *melanocortin receptor 5* (*MC5R*, fc: 1.64, MD20) was upregulated. MC5R is an ACTH and α-melanocyte-stimulating hormone (α-MSH) receptor. α-MSH has an anorexic effect and promotes a negative energy balance. Further agonist of the MC3R and MC4R receptors led to increased food intake. The MC3R and MC4R receptors are mainly expressed in the brain [214]. Islam and Pestka (2003) reported significantly higher adrenocorticotropic hormone (ACTH) levels in the plasma of mice after DON injections [215]. Reduced levels of the MC5R agonist, probably a melanocortin (ACTH, γ-/α-MSH), might have caused the reduced food intake and the increased MC5R expression observed in this experiment. In HEK293, it has been shown that α-MSH stimulation leads to increased ERK1/2 phosphorylation over MC5R, PI3K (*PIK3R1*, fc: -2.12, MD20, D20 vs. MD20), and SNRPE (fc: -1.23, MD20) signaling pathway [216]. Interestingly, Moon and Pestka (2002) showed increased phosphorylation of ERK1/2 in RAW264.7 cells 15 minutes after the addition of DON [217]. To what extent this mechanism is activated remains unclear, but significant regulation of *MC5R*, *PIK3R1,* and *SNRPE* indicates that DON leads to an imbalance of this signal cascade. Wu *et al.* (2006) reported increased spleen weights in rats administered with desacetyl α-MSH, which also binds to MC5R [218]. It is unclear how far the decreased splenic weight was influenced by α-MSH, but the reduced amount of this melanocortin would fit to the increased expression of *MC5R* and the decreased spleen weight.

Another gene related to altered hormone synthesis is the dopa decarboxylase (*DDC*, fc: -1.36, MD20), which performs the conversion of levo-dopa and 5-hydroxytryptophan to dopamine and serotonin [219]. In non-neuronal tissue, it seems to act as a non-specific decarboxylating enzyme and may have other functions. Swamy *et al.* (2004) reported that broiler chickens fed a DON (9.3-9.7 mg/kg feed) and a fusaric acid (12.2-31.9 mg/kg feed) contaminated diet had linearly increased concentrations of serotonin and 5-hydroxyindoleacetic acid (5HIAA) in the pons and linear increased concentrations of serotonin in the cortex. Norepinephrine, a product of dopamine, and 3-methoxy-4-hydroxyphenylethyleneglycol, a product of norepinephrine, were linearly increased in the pons [166]. The downregulation of *DDC* could be the cause of the linear increase of serotonin over a possible feedback mechanism. GTP cyclohydroxylase 1 (*GCH1*, fc: 1.38, MD20, D20 vsMD20) catalyzes the conversion of GTP to dihydroneopterin triphosphate, which is subsequently metabolized to tetrahydrobiopterin (BH4). BH4 is used as an essential cofactor for the serotonin synthesis and the loss of function of GCH1 leads to a deficiency of this neurotransmitter [220]. GCHFR (-1.26, D20 vs. MD20) complexes with GCH1 and controls its activity. BH4 mediates its feedback inhibition of GCH1 activity over the GCHFR protein [9]. Previously, Swamy *et al.* (2004) indicated that the observed feed refusal of the broiler chickens due to DON-contaminated feed might be caused by the increased serotonergic neurotransmitter. In our study, the total feed consumption was significantly reduced in D20 as well as MD20. MC5R, DDC, GCH1, and GCHR seem to be target genes for the study of altered hormone levels in animals fed deoxynivalenol.

Low-density lipoprotein uptake and cholesterol synthesis

Cellular cholesterol levels are related to uptake, efflux, and endogenous synthesis [221]. In our study, several genes were related to cholesterol homeostasis in MD20 and the comparison between D20 and MD20, showing a two-sided confirmation. The low-density lipoprotein (LDL) receptor (LDLR, fc: -1.70, MD20, D20 vs. MD20) is central to maintaining plasma cholesterol levels. LDLR is involved in receptor-mediated endocytosis and the subsequent clearance of cholesterol. Cholesterol is mainly synthesized in the liver and packed into serum lipoproteins, such as LDL [221]. Due to its expression change, LDL-cholesterol levels were measured in heparinized blood plasma and a significant increase of 22.2 % was observed in MD20. Also the fraction of LDL-cholesterol related to total cholesterol was significantly increased by 22%. It seems that the reduced expression of LDLR led to a reduced clearance of LDL-cholesterol and its increase in blood plasma. Vargas *et al.* (2009) showed in HepG2 cells that LDLR mRNA is stabilized by JNK (MAPK8, fc: -1.21, MD20) and that LDL-binding activity was significantly enhanced [222]. The reduced expression of MAPK8 would explain the connection between DON and the reduced LDLR expression with the subsequent increased LDL-cholesterol concentration. The myosin regulatory light chain interacting protein (MYLIP, fc: -1.32, MD20, D20 vs. MD20) is an E3 ubiquitin ligase that triggers ubiquitination of LDLR, thereby targeting the LDLR for degradation. The exposure to extracellular cholesterol increased the expression of MYLIP in vitro [221]. This would indicate that the numerically reduced cholesterol amount might have influenced the reduced expression. In addition, the LDLR amount on the level of proteins does not have to be compellingly reduced, due to a lower degradation. LDLR is negatively regulated by the peroxisome proliferator-activated receptor gamma, coactivator 1 alpha (PPARGC1A, fc: 2.40, MD20, D20 vs. MD20). Overexpression of PPARGC1A led to decreased protein amounts of LDLR in HepG2 cells [223]. PPARGC1A is related to insulin signaling and gluconeogenesis discussed in the corresponding section. Apolipoprotein B (APOB, MD20, D20 vs. MD20) was significantly upregulated in the comparison D20 vs. MD20 (fc: 2.16) and in the qPCR experiment in group MD (fc: 1.24) and D20 vs. MD20 (fc: 1.22). APOB is part of the LDL shell and is essential for the assembly, secretion, and intravascular transport of lipoproteins. APOB contains the domain for the interaction with LDLR. The binding of APOB to the LDLR leads to the clearance of the APOB-containing lipoproteins by endocytosis. Reduced expression of APOB has been related to reduced LDL-cholesterol concentrations in mice [224]. The endocytosis of APOB starts with the binding to the LDLR, which is initially localized at the plasma membrane. The clathrins (*CLTA*, fc: -1.21, D20 vs. MD20) are an important mediator of the endocytosis and the formation of coated vesicles. Golgi associated PDZ and coiled-coil motif containing (*GOPC*, fc: 1.38, D20 vs. MD20) seems to selectively enhance HDL-cholesterol ester uptake into hepatic cells [225]. Other genes related to the endocytosis were *EGFR* (1.49, D20 vs. MD20), *MET* (1.28, D20 vs. MD20) and a gene *similar to receptor tyrosine kinase flk-1/VEGFR-2* (*LOC422316*, fc: 1.35, D20 vs. MD20) according to the KEGG database. Several other genes related to the endocytosis are listed in table 9.1 and 9.2 under the section "Endocytosis." Besides the reduced *LDLR* expression, *APOB* might be responsible for the increase in LDL-cholesterol because total cholesterol was not altered and MD20 showed the lowest cholesterol concentration in terms of tendency. Therefore, we assume that a shift of cholesterol towards the LDL group from other lipoproteins occurred because several genes related to cholesterol biosynthesis showed a reduced expression. An inverse relationship between hepatic lipase (*LIPC*, fc: 1.54, D20 vs. MD20) mRNA and the cell cholesterol content has been described [226]. LIPC clears plasma cholesterol by lipolytic processing and the bridging function, which removes APOB-containing lipoproteins. The clearance of LDL was observed by LIPC in LDLR-deficient mice [227].

The cholesterol concentration in our experiment was not significantly altered; nevertheless, 13 of 23 enzymes responsible for the cholesterol biosynthesis were downregulated. The remaining 10 genes were not present on the microarray or did not have an altered expression. The alteration of this cascade is limited to the expression change between control group and MD20. An increased uptake of cholesterol leads to the repression of the *3-hydroxy-3-methylglutaryl-Coenzyme A reductase* (*HMGCR*, fc: -1.68, MD20). HMGCR is the rate limiting enzyme in cholesterol biosynthesis (fig. 4.6). It has been shown that increased dietary cholesterol leads to the downregulation of *LDLR* and *HMGCR* in the liver of mice [228]. ATP citrate lyase (*ACLY*, fc: -1.21, MD20) is responsible for the formation of acetyl-CoA, which is an essential building block for cholesterol and triglycerides [229]. Acetoacetyl-CoA synthetase (*AACS*, fc: -2.25, MD20, D20 vs. MD20) activates acetoacetate to its coenzyme A ester, which is the starting substance for cholesterol synthesis [230]. *MVK*, *MVD* and *IDI1* were also downregulated in MD20 and are the downstream enzymes of HMGCR in the cholesterol biosynthesis. Mitchell and Avigan (1981) showed a decreased activity of mevalonate kinase (*MVK*, fc: -1.24, MD20) and mevalonate (diphospho) decarboxylase (*MVD*, fc: -1.43, D20 vs. MD20) in the liver of rats fed with a 1% cholesterol supplemented diet [231]. The isopentenyl-diphosphate delta isomerase 1 (*IDI1*, fc: -1.58, MD20) converts isopentenyl diphosphate to dimethylallyl diphosphate [232]. Methylmalonic aciduria cblB type (*MMAB*, fc: -1.40, MD20) catalyzes the formation of adenosylcobalamin, which is important in the catabolism of cholesterol. In human hepatocytes, it has been shown that the downregulation of *MMAB*, due to a single nucleotide polymorphism, was correlated with increased high-density lipoprotein cholesterol [233]. Maxwell *et al.* (2003) measured the gene expression in the liver of mice fed with a high-cholesterol diet. Total cholesterol was increased in the liver, leading to decreased gene expression of *FDFT1*, *NSDHL*, *IDI1*, *SC4MOL*, and *HMGCR* [228], which were also downregulated in our experiment. The farnesyl-diphosphate farnesyltransferase 1 (*FDFT1*, fc: -2.94, MD20) catalyzes the conversion of two molecules farnesyl pyrophosphate to pre-squalene. FDFT1 inhibitors have shown a lowering of cholesterol concentration and *FDFT1* expression is regulated by SREBF2 (fc: -1.42, MD20) and F1 [234]. The lanosterol synthase (*LSS*, fc: -1.34, MD20) is a downstream enzyme of FDFT1. LSS catalyzes the conversion of 2,3-monoepoxysqualene to lanosterol and seems to be an interesting target for the cholesterol synthesis inhibition. Dang *et al.* (2009) showed that a high-fat diet downregulated *LSS* and increased the expression of genes responsible for the reverse cholesterol transport from the liver to the plasma [176]. Those genes were not changed in any of our groups, indicating that only the uptake and the biosynthesis of cholesterol are altered in the liver. The 24-dehydrocholesterol reductase (*DHCR24*, fc: -1.42, MD20) catalyzes the reduction of the Δ24 double bond in the cholesterol biosynthesis. In mouse brains, a reduction of DHCR24 led to reduced membrane cholesterol levels [235]. Waterham *et al.* (2003) reported that lamin B receptor (*LBR*, fc: -1.21, MD20) encodes the sterol Δ14 - reductase. In the human cholesterol biosynthesis pathway, this step is performed by TM7SF2 [236], which is not known in the chicken genome and seems to be replaced by LBR. Sterol-C4-methyl oxidase-like (*SC4MOL* fc: -1.46, MD20) catalyzes the 14-demethyllanosterol to 4alpha-carboxy-4beta-methyl-5alpha-cholesta-8,24-dien-3beta-ol step and its expression is regulated over SREBF2 and the LDL concentration in smooth muscle cells [237]. NAD(P) dependent steroid dehydrogenase-like (*NSDHL*, fc: -1.83, MD20) converts 4-methylzymosterolcarboxylate to 3-keto-4-methylzymosterol [238]. Hydroxysteroid (17-beta) dehydrogenase 7 (*HSD17B7*, fc: -1.81, MD20) catalyses the 3-keto-4-methylzymosterol to 4-

methylzymosterol [239]. In the last step of cholesterol biosynthesis, 7-dehydrocholesterol reductase (*DHCR7*, fc: -1.74, MD20) performs the conversion of 7-dehydrocholesterol to cholesterol [238]. The high number of genes within the same pathway indicates that the biosynthesis, as well as the uptake of cholesterol into hepatocyte, is reduced.

The sterol regulatory element binding transcription factor 2 (*SREBF2*, fc: -1.42, MD20) is mainly related to the cholesterol biosynthesis in contrast to SREBF1, which influences the fatty acid synthesis [176]. The transcription factor analysis showed that both genes influence the gene expression in MD20 (table 4.5). SREBF2 is bound to the membrane and low cholesterol leads to its release and its transcription factor activity. In mice livers, the expression and the promoter binding of SREBF2 was increased when cholesterol synthesis and uptake inhibitor were added to the feed. The expression of *HMGCR* and *LDLR* were also significantly increased as was the binding of SREBF2 to its promoters. Fasting and refeeding had only modest effects on the expression of *SREBF2* in the livers of mice [240]. Wang *et al.* (2009) showed in hepatocyte challenged with palmitic acid a parallel upregulation of *SREBF2, LDLR* and *HMGCR* and an increase in intracellular total cholesterol [241].

Cardiovascular disease and atherosclerosis are related to high levels of plasma cholesterol as is Alzheimer's disease (AD). It has been shown that statins, which reduce serum cholesterol levels, also protect against AD. The cleavage of β-amyloid precursor protein (APP, unchanged) generates the amyloid β-peptide (Aβ), which can accumulate in Aβ deposits and cerebral blood vessels [242]. The *APP cleavage enzyme beta-site APP-cleaving enzyme 2* (*BACE2*, fc: 1.24, MD20) was increased in the liver and shows a possible connection to AD.

Oxysterol binding protein-like 2 (*OSBPL2*, fc -1.24, MD20) is related to the transfer of cholesterol between membranes within the cell [243]. Oxysterol binding protein-like 10 (*OSBPL10*, fc -1.30, MD20) seems to suppress hepatic lipogenesis and very-low-density lipoprotein production. Silencing of OSBPL10 in the human hepatoma cell line led to an increase of apolipoprotein B (*APOB*, fc: 2.16, MD20, D20 vs. MD20) in the growth medium [244]. The stearoyl-CoA desaturase (*SCD*, fc: -1.91, MD20) is a Δ-9 desaturase and responsible for the biosynthesis of oleate and palmitoleate. A disruption in this gene leads to very low levels of VLDL and impaired triglyceride and cholesterol ester biosynthesis [245]. The hydroxy-delta-5-steroid dehydrogenase, 3 beta- and steroid (*HSD3B7*, fc: -1.42, MD20) is related to the primary bile acid biosynthesis from cholesterol by the conversion of 7α-hydroxycholesterol to 7α-dihydroxy-4-cholesten-3-one. According to Schwarz *et al.* (2000) HSD3B7 seems to be the only enzyme that can perform the mentioned reaction and is therefore essential in the bile acid pathway [246]. The Acyl-Coenzyme A oxidase 2 (*ACOX2*, -1.25, MD20, D20 vs. MD20) degrades bile acid intermediates beside other long-branched fatty acids [247]. The changes of the genes from the cholesterol biosynthesis pathway might also have influenced the bile acid pathway over the cholesterol concentration.

It cannot be excluded that DON influenced the expression changes of the cholesterol-related genes by binding or inhibiting transcription factors, signal transducers, receptors, or other proteins that regulate the mentioned genes. *PPARGC1A,* in particular, seems to be a candidate gene over the *LDLR* regulation and in connection to the insulin signaling pathway. The alterations in the cholesterol concentration were only numerical and did not reach the significance level. The changes do correspond to the changes in the gene expression analysis with the decreased expression and content in MD20 compared to control and D20. The cholesterol and LDL-cholesterol concentrations were similar to those in other publications [248, 249]. Chowdhury and Smith reported in laying hens, fed 12.1 mg/kg DON, a reduced-cholesterol concentration after week 4. In the following weeks (8 and 12), significance was no longer attained [250]. In a turkey study, the supplementation of the same mycotoxin absorber (glucomannan polymer) as in our study to a mycotoxin-contaminated diet, showed a significantly reduced serum cholesterol concentration. The authors assumed that the mycotoxin absorber bound bile salts in the intestinal lumen, thereby preventing the enterohepatic recycling and leading to increased hepatic cholesterol synthesis [172]. The reduction of the cholesterol biosynthesis enzyme of our study would, to the contrary, indicate that there is no increase of the cholesterol biosynthesis. The extent to which the reduction of the bile acid biosynthesis responsible genes is caused by a possible bile acid absorption of the mycotoxin absorber is unclear. Gallaher *et al.* (2000) showed a reduced liver cholesterol concentration and reduced cholesterol absorption in glucomannan (Konjac mannan) fed rats [251]. DON seems to influence the LDL-cholesterol concentration over *PPARGC1A*. The biosynthesis of cholesterol, in contrast, could be influenced by the myctoxin-absorber mediated absorption of bile acid.

All the discussed genes in this section were changed in MD20, the comparison between D20 and MD20, or in both tables. The high number of genes related to cholesterol homeostasis indicates that effect of DON and the myctoxin-absorber in MD20 is respectable in contrast to the absence of those genes in D20.

Figure 4.6: Steroid biosynthesis

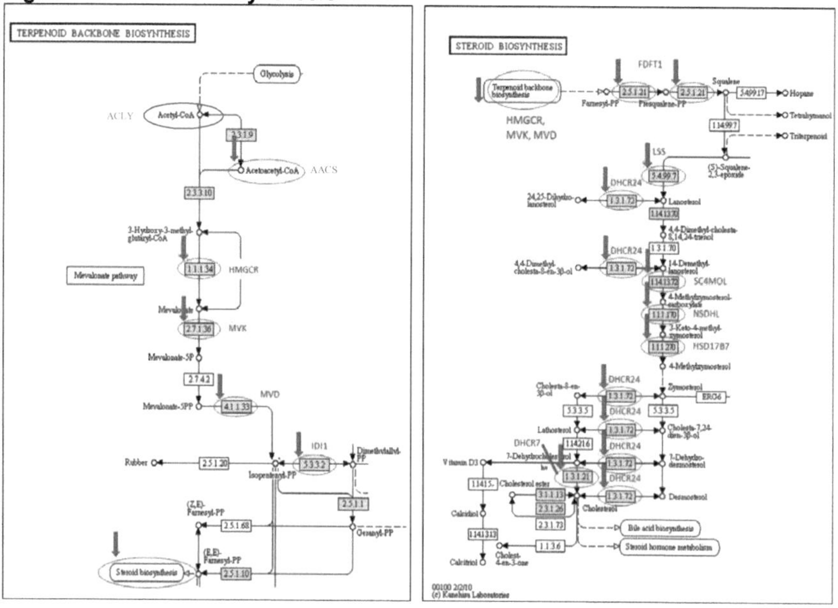

All the downregulated genes are indicated in red. According to the literature, HMGCR is the rate limiting enzyme in cholesterol biosynthesis, which is found in the terpenoid backbone biosynthesis pathway from the KEGG database.

Carnitine acyltranferase

Carnitine palmitoyltransferase 1A (*CPT1A*, fc: 2.55, MD20) expression was significantly upregulated in the qPCR experiment and in the microarray experiment in the comparison D20 vs. MD20. CPT1A is a mitochondrial acetyltransferase, carnitine O-octanoyltransferase (*CROT*, fc: 1.38, D20 vs. MD20) is related to peroxisomes and carnitine acetyltransferase (*CRAT*, fc: 1.72, MD20, D20 vs. MD20) is related to mitochondria and peroxisomes. CPT1A and CROT transesterify medium and long fatty acyl chains and CRAT transesterifies short acyl chains [252]. Carnitine acyltransferases are important for energy homeostasis and fat metabolism and they modulate the pool of acetyl-CoA [253], which is also used for cholesterol biosynthesis. The uptake of long-chain fatty acids is facilitated through the replacement of acetyl-CoA with carnitine by CPT1A. CRAT can modify the acetyl-CoA/CoA ratio and a buffer is formed when acetyl-CoA concentration are high. The transesterification by CROT leads to improved transport of medium-chain acylcarnitins from the peroxisomes to the cytoplasma and the mitochondria [253]. To what extent the expression changes of the carnitine acyltransferases influenced the cholesterol biosynthesis over the acetyl-CoA concentration remains unclear. The high fold change of the carnitine acyltransferase and the coordinated regulation indicate that DON also modifies the energy homeostasis.

Insulin signaling and gluconeogeneis

Glucose production for the peripheral tissue occurs primarily in the liver and is tightly regulated by the insulin signaling pathway. At first, glucose is produced from the breakdown of glycogen stores. Gluconeogenesis is the primary source of glucose with longer fastening [254]. Several genes were found related to gluconeogenesis activation and insulin signaling. The plasma glucose concentration did not show any significant difference among the groups, although the highest concentration was observed in D20, followed by the control and MD20. The highest concentration difference was observed between D20 and MD20. This corresponds to the observation that most of the gluconeogenesis related genes were significantly regulated between D20 and MD20, followed by genes from MD20. Gluconeogenesis is performed by the expression alteration of mainly three enzymes: glucose-6-phosphatase, fructose-1,6-biphosphatase (both unchanged), and phosphoenolpyruvate carboxykinase 1 (*PCK1*, fc: 7.76, MD20, D20 vs. MD20). PCK1 is the rate-limiting enzyme in gluconeogenesis and its production is controlled at the level of transcription [254]. The promoter of PCK1 seems to integrate nutritional signals mediated by the transcription inducers cAMP and glucocorticoid hormones and by the inhibitor insulin. It influences the blood glucose level and its overexpression is related to obesity [255]. Forkehead box O1 (FOXO1) and PPARGC1A (fc: +2.49, MD20, D20 vs MD20) cooperate to stimulate the expression of *PCK1* [256]. FOXO1 (unchanged) has been detected in the transcription factor analysis (table 4.5).

FOXO1 and PPARGC1A are both targets of insulin signaling and can activate gluconeogenesis. FOXO1 binds directly to the promoters of gluconeogenic genes and is directly phosphorylated by AKT, which, in turn, is activated over the cascade starting at the insulin receptor (*INSR*, fc: 1.52, D20 vs. MD20) over the phosphoinositide-3-kinase (*PIK3R1,* fc: -2.12, MD20, D20 vs. MD20) and PDPK1 (unchanged). The ability of PPARGC1A to induce gluconeogenesis depends on the interaction with FOXO1. The ability of insulin to suppress the PPARGC1A-mediated gluconeogenesis depends of the insulin pathway to modify FOXO1. PPARGC1A can also regulate the transcription of PCK1 through the hepatocyte nuclear factor 4 alpha (*HNF4A*, fc: -2.35, D20 vs. MD20). Further *PCK1* expression is regulated by c-Jun, HNF1A and ESR1, which were also detected in the transcription factor analysis (table 4.5) and peroxisome proliferator-activated receptor alpha (*PPARA*, fc: 1.61, D20 vs. MD20) [257].

Taniguchi *et al.* (2006) have shown that INSR, PI3K heterodimer, and AKT are the three best critical nodes for the insulin-signaling pathway [257]. The growth factor receptor-bound protein 10 (*GRB10*, fc: -1.30, MD20) downregulates the INSR function beside other inhibitors [257]. Therefore, the downregulation of *GRB10* might lead to an increase of the INSR function. The GRB2-associated binding protein 1 (*GAB1*, fc: 1.20, D20 vs. MD20) is a substrate of the INSR. The main substrates of the INSR are IRS1 and IRS2, which are available on the microarray, but they were both expressed below the set threshold of the 63.4 percentile. PI3K consists of a regulatory (PIK3R1) and a catalytic subunit. The activation of the catalytic subunit depends on the interaction with the regulatory subunit. This activation depends on the binding of a phosphorylated insulin receptor substrate from the INSR. PIK3R1 encodes for 65%-75% of the regulatory subunit. According to the review of Taniguchi *et al.* (2006), the knockdown of PIK3R1 led to increased insulin sensitivity. It has also been shown that PIK3R1 is required for the insulin-stimulated activation of JNK (*MAPK8*, fc: -1.21, MD20), which is related to the stress-kinase pathway. AKT is the downstream target of PI3K. Akt phosphorylates FOXO1 and causes thereby the insulin-mediated FOXO1 inhibition. FOXO1 activates gluconeogenesis in the AKT-unphosphorylated form. AKT activity is also controlled by several inhibitory molecules like protein-phosphatase 2A [257]. The PP2R4 (fc: 1.29, D20 vs, MD20) and PPP2R2D (fc: 1.24, D20 vs. MD20) are subunits of the PP2A complex, which is a negative regulator of AKT phosphorylation. The expression of *the serum/glucocorticoid regulated kinase 2* (*SGK*2, fc: 1.48, MD20) was upregulated in the qPCR experiment and downregulated between D20 and MD20 in the microarray experiment. SGK2 is phosphorylated by PDPK1 and dephosphorylated by PP2A. A partial inhibition of the SGK2 phosphorylation could be attained by the preincubation with PI3K-inhibitors after the stimulation with H_2O_2 [258]. Other possible signaling nodes are GRB2 (fc: -1.27, D20 vs. MD20) and SOS1 (fc: 1.24, MD20, D20 vs. MD20), which are activated by the insulin receptor substrates, like GAB1. Insulin has been shown to activate JNK1 (*MAPK8*, fc: -1.21, MD20) [257]. Gluconeogenesis can also be promoted by the inhibition of the pyruvate dehydrogenase complex (PDC). The pyruvate dehydrogenase kinase isozyme 4 (*PDK4*, fc: 2.47, D20 vs. MD20) can phosphorylate PDC and thereby induce a shift of pyruvate from the fatty acid synthesis toward the gluconeogenesis. Aquaporin 9 (*AQP9*, fc: -1.27, D20, D20 vs. MD20) transports glycerol into the liver, which is a substrate for gluconeogenesis. *AQP9* is transcriptionally regulated by CCAAT/enhancer binding protein beta (*CEBPB*, fc: -1.45, MD20, D20 vs. MD20), HNF4A, signal transducer, and activator of transcription 1 (*STAT1*, fc: -1.29, MD20, D20 vs. MD20) and FOXO1.

FOXO1 has the ability to increase the expression of *insulin-like growth factor binding protein 1* (*IGFBP1*, fc: 2.54, D20 vs. MD20) in fasted mice [254]. Maresca *et al.* (2002) reported in DON challenged HT-29-D4 cells a decrease of deoxyglucose uptake, which is transported by all GLUT family members like the solute carrier family 2 member 2 (*SLC2A2*, fc: -1.34, MD20) [259]. The downregulation of *SLC2A2* might be an indication that the glucose uptake is also reduced in regard to the possible increase of the gluconeogenesis. Awad *et al.* (2004) showed that glucose uptake from the lumen was decreased in broilers fed up to 9.5 mg/kg DON [177]. In laying hens, a significant increase of glucose plasma concentration was observed between the group with DON-contaminated wheat (12.1mg/kg DON) and the group with contaminated and glucomannan polymer supplemented diet [250]. This is in contrast to our results, which showed a tendency to have a decrease in glucose concentration between the control and MD20. The glucose concentration did increase slightly between the control and D20, but without any indications from the gene expression of gluconeogenesis and insulin signaling related genes.

Zhou *et al.* (2005) showed that DON can induce AKT phosphorylation with a subsequent increase of phosphorylated FOXO1 [150]. Our hypothesis is that after each feed uptake the signal transducer AKT seems to be phosphorylated. This might have led to the deregulation of insulin signal transduction. DON seems to imitate insulin over the phosphorylated AKT, which, in turn, blocks gluconeogenesis.

To counterbalance the insulin-unstimulated block of the gluconeogenesis, several discussed genes showed an altered expression. Regarding the experimental design and the slaughter, the expression changes of the insulin signal transduction cascade cannot be explained by fastening. The feed was removed just before the slaughter, which was performed in the same building. Moreover, the stomachs and the intestines of the birds were well filled when samples were taken. Therefore, the observed effects seem to be caused by DON and the addition of the mycotoxin absorber.

Figure 4.7: Insulin signaling

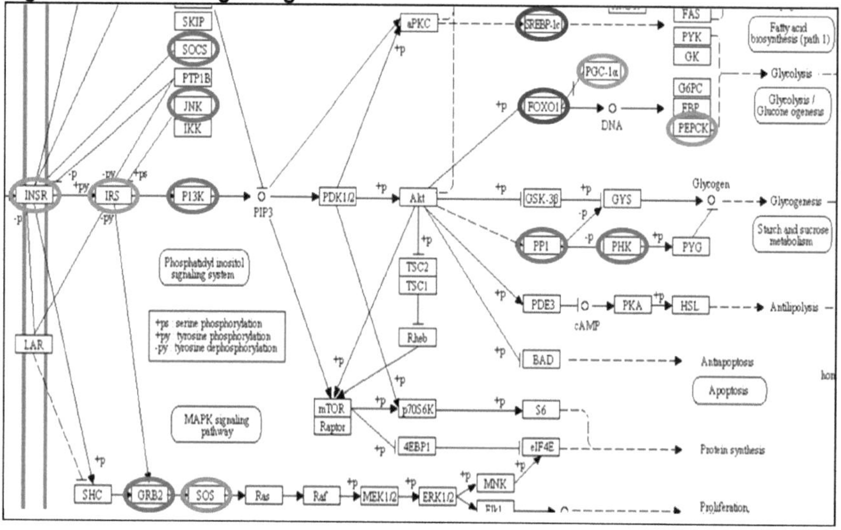

This pathway from the KEGG database shows the insulin signaling for the activation of gluconeogenesis.

Several objects in the insulin signaling pathway are complexes and groups of genes are summarized to one term. The following list indicates the terms on the map and the corresponding genes with an altered expression: INSR: Insulin receptor, IRS: GAB1, JNK: MAPK8, SOCS: SOCS2 PI3K: PIK3R1, FOXO1, PGC-1α: PPARGC1A, PEPCK: PCK1, GRB2, SOS: SOS1, PHK: PHKA2, and PP1: PPP1R3B and PPP1R3C, SREBP-1c: SREBF1.

Red: downregulated, green: upregulated, blue: transcription factor analysis

Nutrient transport

In our previous report, other genes like the fructose transporter *SLC2A5*, the palmitate transporter *SLC27A4*, the D-serine transporter *SLC7A10,* and the neutral amino acid transporter *SLC7A5* were significantly changed in the livers of 23 day old broilers fed a DON-contaminated diet with up to 5 mg/kg DON (Deoxynivalenol alters the gene expression in the liver and the jejunum at low contamination levels). In the current experiment, only the gene *SLC7A5* (fc:-1.23, D20 vs. MD20), a transporter of neutral amino acids with branched or aromatic side chains, was significantly regulated. SLC7A5 transports amino acids like leucine, isoleucine, phenylalanine, methionine, tyrosine, histidine, tryptophan, and valine [260]. SLC38A2 (fc: 1.32, D20) is neutral amino acid transporter and its expression is induced due to amino acid starvation in HeLa cells [261]. In an incorporation study with mice, the ^{14}C leucine incorporation into the liver was decreased at 20mg DON/kg feed. It was assumed that DON inhibited the protein synthesis and therefore the reduction of the ^{14}C leucine amount occurred [262]. It seems that the effect of DON on nutrient absorption might be age-dependent similar to the decreased body weight of the broilers after the second and third weeks.

Immune response

The toll-like receptor 5 (*TLR5*, fc: -1.64, D20 vs. MD20) belongs to the innate immune response genes and recognizes pathogen-associated molecular patterns. In murine macrophage, DON stimulation led to increased expression of the cytokine interleukin 1 beta (IL1B) [263]. Hayashi *et al.* (2003) showed that IL1B is induced when neutrophils were treated wtaith a TLR5 agonist [264]. In the current experiment, the *IL1B* (fc: -1.37, MD20) showed a reduced expression. Pestka and Zhou (2006) showed that priming of the macrophage with the TLR5 agonist flagellin, highly increased *IL1B* mRNA after the DON treatment. The TLR5 agonist was capable of sensitizing the cells to the DON-induced expression increase of *IL1B*, which was much higher than without the priming [263]. The downregulation of the *TLR5 – IL1B* might be a part of a possible explanation for the suppression and the desensitization of the immune system by DON in broilers.
The *suppressors of cytokine signaling* (*SOCS2*, fc: -1.48, D20, fc: -1.72 MD20) was downregulated also in the qPCR experiment. SOCS proteins are critical in the negative regulation of growth factor and cytokine signaling pathways. *SOCS2* expression is regulated by the poly (ADP-ribose) polymerase family, member 1 (PARP1, fc: -1.23, MD20). It has been reported that DON can rapidly upregulate cytokines like interleukin-6 *in vivo.* It has also been predicted that DON binds ribosomes, leading to the phosphorylation of MAPK14 (unchanged, dephosphorylated by DUSP1) and subsequently to increased mRNA stability and increased translation of cytokine mRNA [194]. By a different normalization method (MAS5), the *cytokine inducible SH2-containing protein* (*CISH*, fc: -1.40, D20) was significantly downregulated. Amuzie *et al.* (2009) showed in the liver of mice, treated with a single oral DON dose, an increased expression of the cytokines TNF-α and IL-6 with a subsequent upregulation of *cytokine inducible SH2-containing protein* (*CISH*), *SOCS2* and *3*, and the downregulation in the liver of *insulin-like growth factor binding protein, acid labile subunit* (*IGALS*, fc: -1.22, D20 vs. MD20). The increase of *SOCS2* expression was affected least compared to the other genes and CISH and SOCS3 reached peak concentration one hour after the TNF-α expression peak. *IGFALS* expression was decreased maximal at the end of the experiment (5 hours). In addition, IGFALS has been related to growth hormone signaling and growth reduction [265]. The difference to our results, which showed a decreased *SOCS2* and *CISH* expression, indicates that chronic DON exposure might reverse the effect of DON on the expression of *CISH* and *SOCS2*. It has been shown that SOCS2-deficient mice exhibit increased expression of *peroxisome proliferators-activated receptor-gamma coactivator 1 alpha* (*PPARGC1A*), which was also upregulated in our experiment (fc: +2.40, MD20). Furthermore, SOCS2 deficiency in mice caused gigantism associated with muscle *PPARGC1A* upregulation [266]. This connection seems to be interesting because *PPARGC1A* expression change was also discussed in connection with the LDL-cholesterol and gluconeogenesis related genes.
c-Jun was detected in the transcription factor analysis of significantly changed genes in MD20 (table 4.5). Wong *et al.* (2002) showed in nuclear extracts of RAW 264.7 murine macorphage an increased binding of c-Jun and phosphorylated c-Jun in supershift electrophoretic mobility shift

assay, when cells were treated with DON. Similar effects were observed for CEBPB (fc: -1.45, MD20, D20 vs. MD20), which was also found as mediator of the cytokine superinduction. These results proved *in vitro* that these transcription factors had increased binding activity to the consensus sequence upon DON stimulation. c-Jun is mainly phosphorylated by JNK (*MAPK8, JNK1*, fc: -1.21, MD20) [267]. Beside the altered expression, CEBPB was also detected in the transcription factor analysis with the genes from MD20 and the comparison D20 vs. MD20.
According to Pestka *et al.* (2004) CREB1 binding increased upon DON administration [194], which was detected in the transcription factor analysis of MD20 and D20 vs. MD20 (table 4.5). This indicates that similar signaling effects occur in the liver of broilers. The downregulation of *IL1B, CEBPB,* and *JNK1* further indicates that chronic DON administration leads to opposite effects observed in acute state. Possibly a recurrent stimulus might be counteracted by reducing the amount of signal transducers. This seems be an approach to explain the immune suppressive effect of DON.

Signaling

DUSP1 (fc: 1.36, MD20) is a mitogen-activated protein kinase phosphatase and acts as a negative feedback regulator of MAPK signaling by dephosphorylating, especially JNK1/2 (*MAPK8, JNK1*, fc: -1.21, MD20), MAPK14 (p38), and to lesser extend ERK1/2. An expression and a protein amount increase of DUSP1 have been shown in HepG1 cells treated with DON. The use of DUSP1 knock-down cells showed an increased activation of ERK1/2 and MAPK14, and the phosphorylation of JNK1/2 was prolonged. This led to the increase of the cleaved apoptosis marker PARP1 (fc: -1.23, MD20) in the absence of DUSP1 and the inactivation of JNK1/2 [268]. *PARP1* was upregulated in the previous experiment with up to 5 mg/kg DON. DUSP1 seems to be an important MAPK signaling regulator to prevent apoptosis in the liver. DUSP16 (fc: +2.15, MD20, D20 vs. MD20) can be phosphorylated by MAPK14 and seems to dephosporylate JNK1/2 and MAPK14. DUSP14 (fc: -1.50, MD20, D20 vs. MD20) mainly dephosphorylate ERK1/2 and JNK1/2 [269]. The chronic administration of DON could have led to a desensitizing of the ERK1/2, JNK1/2, and MAPK14 signaling. MAPK8 is an upstream phosphorylation regulator of PARP1 and there seems to be a feedback loop from PARP1 to MAPK8. PARP1 is also related to DNA repair and cell death process [270].
Li and Pestka (2008) showed in DON-treated RAW 264.7 cells a decreased expression of *ribonuclease L* (*RNASEL*, fc: -1.41, MD20) and a decreased RNase activity after six hours. RNASEL is related to the inhibition of protein synthesis, induction of apoptosis, and antiviral activity. The activation of RNASEL leads to the cleavage of the 28S rRNA and subsequently to cellular stress response and the activation of the JNK1/2 (MAPK8, fc: -1.21, MD20) pathway. It has been suggested that the binding of DON to the 28S rRNA and the subsequent confirmation change of the tertiary structure might be sufficient for the cleavage by RNASEL [167]. RNASEL also downregulates *EIF2AK2* mRNA (unchanged), which is activated by DON in cloned macrophages [194]. In RNASEL knockdown mouse fibroblast lines, the *HuR* mRNA half-life were increased, due to the adenylate-uridylate-rich element (ARE) in the 3' untranslated region [271]. It has already

been shown that DON leads to increased half-life of certain mRNA containing the ARE [194]. It seems that RNASEL is directly involved in EIF2AK2 activation and increasing mRNA stability. Our observation of the downregulated gene *RNASEL* might indicate that the chronic DON administration leads to an adaption of 28S-EIF2AK2 signal cascade by the downregulation of *RANSEL*.

Interferon induced with helicase C domain 1 (*IFIH1*, fc: -1.24, MD20) is responsible for sensing viral single- and double-stranded (ds) RNA and the induction of interferon (IFN). Wang *et al.* (2009) showed that IFIH1 knockdown in human bronchial epithelial cells led to *IFN-β, -γ1, γ-2/3* downregulation, when the cells were challenged with rhinovirus 1B and 39. The toll-like receptor adaptor molecule 1 (*TICAM1*, fc: -1.26, MD20, D20 vs. MD20) is an adaptor protein of TLR3, which is a dsRNA sensing protein. The signal transduction is performed over TICAM1 and IFN expression. Both dsRNA sensing proteins IFIH1 and TLR3, with TICAM1, have the downstream intermediate IRF3 (unchanged), which then regulates IFN expression [272]. Waché *et al.* (2009) showed that the binding of IFN-γ to its receptor (IFNGR) induced Jak1 and Jak2, which are associated with IFN-γR1/R2 (*IFNGR2*, fc: -1.21, MD20). Subsequently, STAT1 (fc: -1.26, MD) is activated [273, 274]. The chronic exposure to DON led to the downregulation of *IFIH1* and *TICAM1*. Reduced level of IFIH1 and TICAM1 might be a reason for the decreased innate immune response in relation to the sensing of viral RNA leading to the observed immune suppressive effect of DON. It has been speculated that due to the translation inhibition of DON the immune suppressive effect occurred. The absence of significantly changed expression of ribosomal proteins, with exception of *MRPL38* (fc: 1.38, MD20) and *MRPS18A* (fc: -1.36, MD20, D20 vs. MD20), contradicts this thesis. We have the hypothesis that the chronic sensing of dsRNA, due to the binding of DON to 28S ribosomal subunit, led to the downregulation of genes related to the immune response. Kouadio *et al.* (2005) showed an increased inhibition of the cell viability in Caco-2 cells, challenged with DON, with the MTT assay. With the MTT assay the conversion of tetrazolium salt to fromazan is measured in mitochondria. Therefore, they concluded that DON targets the mitochondria [78]. This would also explain the significant expression changes of *mitochondrial ribosomal* genes. The indications from the expression analysis show that there seems to be a desensitizing of the signaling cascade upon chronic exposure to DON. In general, most of the genes that were activated in cell culture experiments in the acute phase, due to DON administration, were downregulated in our experiment upon a chronic exposure of the animals. There seem to be mechanisms that lead to an altered expression, when a signaling cascade is repeatedly activated under chronic homeostatic imbalances. This adaption might lead to a recovery from the downstream effects of a stressor, but it might cause a reduced responsiveness towards other stressors like pathogens.

Conclusion

DON led to significantly decreased body weight, weight gain, and feed consumption of the broiler chickens. With exception of the reduced feed consumption over the feeding period, the myctoxin absorber could prevent the birds from the detrimental effects. Toward the end of the feeding trial, the broilers from D20 recovered from the growth depression, showing that an adaption occurred. In D20, the genes from the xenobiotic metabolism were upregulated and the cell cycle regulating genes seem to be altered to prevent a cell cycle arrest in G_2/M phase. Genes related to decreased cell permeability and hormone signaling were mostly regulated in MD20. For the first time, the decreased cholesterol biosynthesis, the increased LDL-cholesterol concentration, and the altered expression of LDL-associated genes have been described as a consequence of DON-contaminated feed and the addition of the mycotoxin absorber. As a possible downstream signal of the AKT-FOXO1 cell survival cascade [150], the genes involved with the insulin signaling and the gluconeogenesis were significantly regulated in the comparison between D20 and MD20. The expression change of genes related to the nutrient transport, the immune response, and the signaling, due to the "ribotoxic stress response," confirm the *in vitro* results observed by other authors. The assignment of the detected genes to cell culture experiments increases the reliability of our results and shows that the experiment is reproducible and comparable to other studies.

5. Summarizing discussion

In all three experiments, the main challenge was to establish an overview about the function of the genes with a significantly altered expression and the assembly of those genes into reasonable biological networks. For the description of the gene function, publications with mainly cell cultures and knock-out animals were used. The studies were selected to be as closely related as possibly to the experimental conditions of this work.

In the first experiment with the feed additive Se-yeast, the reduced oxidative stress caused by the increase of glutathione peroxidase activity in the blood plasma and the increased *GPX* expression in the liver led to several secondary effects. The reduced expression of the ubiquitin proteolytic pathway genes and the increased apparent metabolizable energy in the Se-yeast group indicated a reduced amount of oxidized and damaged proteins. The increased PUFA concentration and the altered expression of PUFA biosynthesis enzymes indicated a reduced usage of PUFA. The reduction of PUFA consumption seems to be related to reduced oxidative stress, which has been indirectly confirmed with the decreased activity and expression of the reactive lipid aldehyde, decomposition enzyme AKR1B1. A better protection of membranes over the expression increase of *GPX4* and the decreased expression of vesicular-transport-related genes might also have contributed to the reduced PUFA usage.

In the second experiment, the feeding of DON contaminated wheat led to altered expression of nutrient transport genes, an increase in single stranded DNA repair genes, and a decrease of mRNA degradation enzymes, probably causing the observed increase of mRNA stability [126]. The binding of DON to the 28S ribosomal RNA seems to have caused the expression deregulation of translation initiation genes. The altered expression of gap and tight junction genes seems to be related to the decreased transepithelial electrical resistance observed in cell cultures [130]. The changes of gene expression show that DON influences the metabolism already at 2.5 mg DON / kg feed, which is below the maximal allowed DON concentration of 5 mg/kg for chicken feed in the European Union [88]. Therefore, a reduction of the maximal allowed concentration in the feed should be further discussed.

In the third experiment with 17.7 mg DON / kg feed, the mycotoxin influenced the body weight and the feed consumption negatively. The mycotoxin-absorber could prevent the reduced body weight and diminish the reduced feed consumption. In the group with 17.7 mg DON / kg feed (D20), the xenobiotic metabolism was influenced. The observed cell cycle arrest in cell cultures [168, 169], seemed to be counterbalanced by the alteration of several G_2/M transition genes. In addition, the increased AME indicated that the development of the birds was delayed. In the group with 17.7 mg DON / kg feed and 0.2% mycotoxin-absorber (MD20), the expression of genes responsible for the monolayer permeability, the hormone signaling, and the immune system showed effects that have been shown in DON contamination experiments. The increase of LDL-cholesterol concentration in MD20 could be explained by alterations on the transcript level. The decrease of 13 of 23 cholesterol biosynthesis enzymes did not have a direct influence on total cholesterol concentration in blood plasma. The down-regulation of *RNASEL* (fc: -1.41, MD20) seems to reduce the cleavage of the 28S RNA, which is caused by the binding of DON and the subsequent conformation change [167].

In the comparison of D20 vs. MD20, the insulin-signaling cascade was deregulated, leading to an increased expression of gluconeogenesis genes *PPARGC1A* and *PCK1*. Even though the mycotoxin-absorber could prevent the detrimental effects of DON, significantly altered expression of gene clusters could be related to DON-related effects, such as the genes responsible for the immune response, signal transduction, and the cleavage of the 28S ribosomal subunit, which have been reported in the literature [11, 126].

Summarization of the three gene expression analysis

To get an overview of the three projects and to describe the relative distance of each sample from the other, a principle component analysis (PCA) was performed (Figure 5.1, additional perspectives: Appendix D, Figure 10.1). The genes for the principle component analysis had to have a gene expression raw value of 62.2th percentile of all genes in 22 of 44 samples. The PCA shows the high variability between the samples and the experiments. The differences between the experiments clearly show that the time point of development is highly important for gene expression. The birds from the first DON experiment were held for 23 days, in contrast to the other experiment with a feeding period of 35 days. The birds forming the first DON experiment have a completely different location, compared to the second DON and the Se-yeast experiments. The difference between the second DON experiment and the Se-yeast trial is based on the feed composition, which was modified to attain the intended DON concentration of 20 mg/kg. The 27.5% corn portion was reduced to 17.7%, with the aim to increase the wheat concentration from 35% to 45%. Another feed premix, and rapeseed oil instead of linseed oil, were used in the second DON experiment, which might also have contributed to the different location of the samples in the PCA. Figure 10.1 shows the PCA of different treatments and the grouping of the experiments. The distribution of the control samples shows a similar variability to the treatment samples. It can be expected that the majority of the average expressions of each gene from the control group is different depending on the experiment. The volcano plot (fc: > 1.2, $p < 0.05$) of uniquely control samples between the first and the second DON experiment (control samples 1st DON exp. vs. control samples 2nd DON exp.) shows 4454 significantly altered genes (30.6% of all expressed genes). 1533 (10.5%) or 4915 (33.7) were significantly regulated between the Se-yeast and the first or the second experiment, respectively. This comparison of the control samples shows that the observable effects of a feed supplement might be different depending on the standard feed mixture used for an experiment. Therefore, it is crucial to plan thoroughly the experimental design for a gene expression analysis. This experiment intended to have a diet that was comparable to commercial feed mixtures. Therefore, a premix was added, resulting for instance, in 30 mg/kg vitamin E, which is also an antioxidant like selenium.

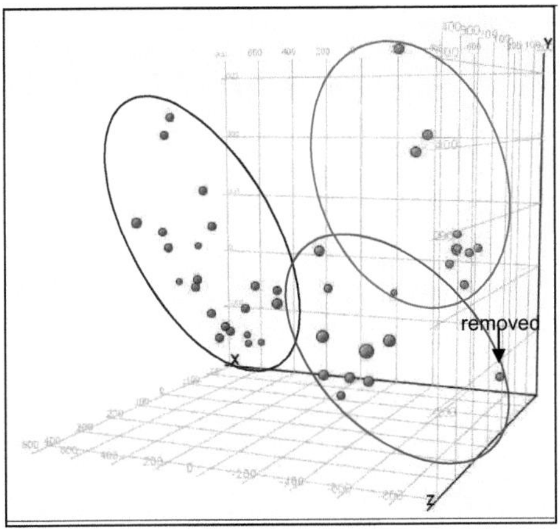

Figure 5.1: Principle component analysis

Principle component analysis, three dimensonal
Samples and circle brown: Se-yeast experiment
Samples and circle red: First DON experiment up to 5mg/kg DON
Samples and circle blue: Second DON experiment with up to 17.7mg/kg DON
The sample with the arrow was removed from gene expression analysis.

The PCA (Figure 5.2 and 5.3) also shows the high variability of the samples within all three groups. One reason could be that the broilers were F2 hybrids with high heterozygosis compared, for instance, to inbred laboratory mice with a high degree of homozygosity. In poultry production, different roosters are used for breeding. Even slightly uneven genetical material from the paternal side or from the maternal side might lead to elevated variance in the gene expression profile. Therefore, the differences in expression fold change were lower, compared to other studies. Afman and Müller concluded that microarray analysis works best when genetic variation is minimal, which is mainly the case in animals with inbred strains but not in hybrids, like broiler chicken, or humans [3]. The results from the PCA with the high variability show that the use of farm animals, especially hybrids, leads to the necessity of having at least five to seven microarrays per group. The low number of significantly altered genes from the first DON experiment seems to be caused by the low number of microarrays.

Figure 5.2: Principal component analysis all treatments, three components

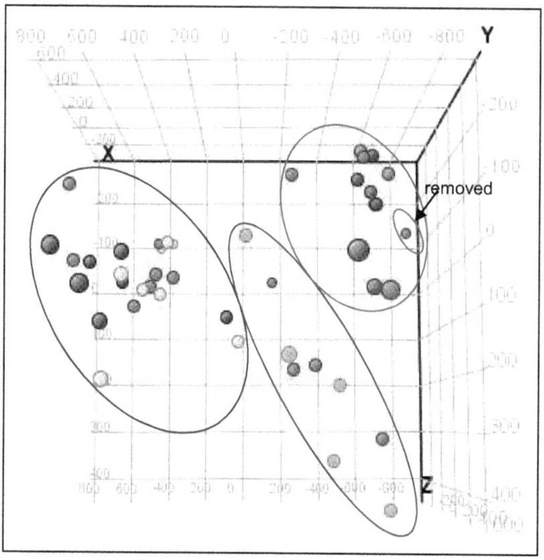

Principle component analysis with all treamtents.

Brown circle: Se-yeast experiment, red: Control samples, light blue: Se-yeast samples

Red circle: First DON experiment, red: Control samples, blue: 1mg/kg DON, brown: 2.5mg/kg DON, green: 5mg/kg DON

Blue circle: Second DON experiment, red: Control samples, grey: 20mg/kg DON, black: 20mg/kg DON + myctoxin absorber

Sample with arrow was removed from gene expression analysis

Figure 5.3: Principal component analysis all treatments, two components

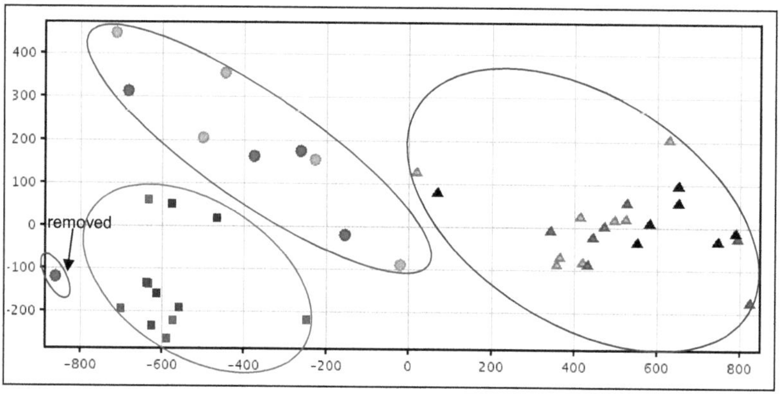

Principle component analysis with all treamtents.

Brown circle: Se-yeast experiment, red: Control samples, light blue: Se-yeast samples

Red circle: First DON experiment, red: Control samples, blue: 1mg/kg DON, brown: 2.5mg/kg DON, green: 5mg/kg DON

Blue circle: Second DON experiment, red: Control samples, grey: 20mg/kg DON, black: 20mg/kg DON + myctoxin absorber

Sample with arrow was removed from gene expression analysis

A smart method to avoid genetic background disturbance is the use of nutrigenetics. With the genotyping of the single nucleotide polymorphisms, each treatment group could be subdivided, depending on the alleles of the animals in one group. For example, the animals could be divided into three subgroups, when a gene contained alleles A/B. The gene expression would then be compared between control AA and treatment AA, or control AB (BB) and treatment AB (BB). For every gene, the subgrouping would be rearranged always to compare the same genotypes. This approach would result in more accurate p-values and fold change results (Figure 5.4).

Figure 5.4: Nutrigenomics in relation to genetic background

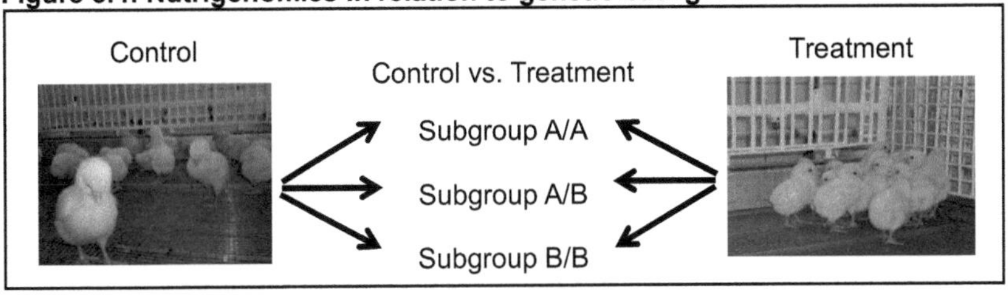

The two groups are subdivided for every gene variant to compare subsequently the gene expression. This would allow a comparison between genetically identical animals.

In addition, the genetic information gained from these measurements might be integrated into breeding programs. For instance, the gene expression of *GPX1* and *4* were significantly increased in the Se-yeast group. Further, the magnitude of the increase was different between the samples. The real-time PCR showed a *GPX1* expression increase of 2.8 up to 8.5 in the Se-yeast group, based on an average of 1 in the control group. The characterization of the phenotype (expression increase of *GPX1* and *4*) clearly shows differences, which could possibly be traced back on an allele. The breeding for high *GPX* expression could be beneficial for nutrient efficiency. This example shows that the phenotyping of the animals for gene expression study would allow researchers in breeding biology to obtain the breakdown of a complex characteristic on a well-defined measurement. The measurement of protein amounts and activities and metabolites, like the fatty acid profile for meat quality, is frequently performed and gene expression analysis could complete the repertoire of analysis.

Perspectives

It seems beneficial to test the relation between Se supplementation and the increased N retention with an N reduced diet. In an N limited diet, the supplementation Se-yeast might increase body weight gain and N-reduced feeds would reduce environmental pollution and costs.

The fatty acid profile was also altered upon Se-yeast supplementation. A detailed measurement of the fatty acid profile in the meat could give valuable information on the meat quality and the possible alteration of particular fatty acids might be beneficial for the health of consumers.

The experiment with up to 5 mg DON / kg feed resulted in significantly altered expression of genes in the liver and the jejunum. With an increased number of animals, an investigation related to nutrient uptake might increase the knowledge about low and chronic doses of DON in the diet of broilers.

The last DON experiment showed altered expressions of cholesterol biosynthesis genes in relation to the mycotoxin-absorber. It would be interesting to perform an experiment with the supplementation of the mycotoxin-absorber to a high cholesterol and/or fat diet. The measurement of the gene expression in the liver, the different lipoprotein cholesterol concentrations, and total cholesterol concentration in the plasma and the liver could explain to what extent the mycotoxin-absorber influences the cholesterol biosynthesis and the LDL-cholesterol concentration. In addition, there was evidence that showed the activation of signal transducers, transcription factors, or activity changes of genes, which are DON-related. The confirmation of those changes in living animals would increase the reliability of previous results and the corresponding genes and clusters of genes could be used as biomarkers for mycotoxin-contaminated diets in the living animals. The gene *RNASEL* seems to be an especially good biomarker. *RNASEL* seems to be a key gene of translation inhibition, because it cleaves the DON-binded 28S ribosomal subunit.

Remarks on the analysis and interpretation of the gene expression analysis

Several implications are described in this work, when the expression of a gene is altered. The expression change of a gene leads to an activity change, which has been shown with AKR1B1. Reduced expression leads to reduced activity, although the activity measurement was not completely specific for AKR1B1, and therefore, the reduction was lower in the Se-yeast group compared to the expression change.

The second possibility is that the expression change does not lead to an alteration of the influenced metabolite. The genes *LDLR* and *APOB*, which are related to the LDL turnover, were significantly regulated in MD20 and D20 vs. MD20. The LDL measurement, however, showed a significant concentration change only in MD20, compared to the control but not for D20 vs. MD20. Other influences, like translational and post-translational modifications, protein degradation and deactivation, might lead to no detectable differences at the protein or metabolite levels.

Another observation showed that the change of the metabolite concentration leads to an altered expression of enzymes, which are involved in the modulation of the corresponding metabolite. The *FADS1* expression was decreased in the Se-yeast group, which led first to the assumption that the product arachidonic acid would have a decreased concentration. The fatty acid analysis showed the exact opposite, that the arachidonic acid concentration in Se-yeast group was significantly increased. Over a feedback loop mechanism, the metabolite influences the gene expression to maintain homeostasis. The gene expression change alone, therefore, cannot always predict in which direction the metabolite concentration is changed. For this reason, verification at the level of protein activity or metabolite concentration needs to be performed.

Real-time PCR cannot always confirm the results from the microarray analysis. Due to the statistical testing of a high number of genes, several genes are detected by chance (false positive). Another error source is that the real-time PCR measurement compares the expression of a target gene with a housekeeping gene, which is expected to be expressed at the same level in all samples.

In contrast, the normalization of the microarray data is performed on the total signal of the expressed genes. The housekeeping genes used in the real-time PCR experiments were chosen, based on the microarray results. The expression of those genes should be similar between the groups and have a low variation between the samples. Several of the housekeeping genes were tested, and the results showed that some work better than others do. This shows that real-time PCR is susceptible to low quality housekeeping genes.

The animals for the microarray analysis were chosen according to their average weight. The additional samples from birds, which are lighter and heavier, might negatively influence the verification rate in the real-time PCR analysis. This can be the result of the genetic background of the birds, which are double hybrids with a higher number of heterocygote gene loci compared to inbred animals. In the group D20 (second DON experiment), the standard error of the body weight was increased, compared to MD20 and the control. This body weight variation might also influence the gene expression and lead to a low number of differently expressed genes.

In the first DON experiment, the correlation between the DON concentration (1, 2.5, 5mg/kg DON) and the gene expression was calculated. For the calculations, all genes with a significant alteration in at least one group were used for the analysis. The results showed that only 16 genes of 566 had a correlation higher than 0.95. In the second DON experiment, the mycotoxin-absorber should have decreased the concentration of bioactive DON in the intestine. The gene expression analysis showed 17 genes that were significantly regulated in D20 and MD20 of 611 total genes. The identification of genes with a correlation analysis, besides the filtering on significance, does not seem to be informative. These results show that the altered concentration of a feed contaminant leads to an altered mode of action and influence to other pathways. Even though parameters like body weight and feed consumption are influenced by concentration changes, which was observed in the second DON experiment with the mycotoxin-absorber.

In the three different studies, several pathways were described with the corresponding genes that showed a significantly altered expression. Depending on the portion of genes detected in the expression analysis compared to the total gene number of a pathway, the reliability of the biological conclusions is high, even though the expression change of the single genes was based on a fold change below two and a significance value below 0.05. Therefore, a classification of the genes to pathways, signal transduction, or similar function is very valuable for reliability and the interpretation of the data.

The addition of a single supplement or contaminant cannot be examined independently because of interactions to other nutritional components. The effect of an additive might be different, depending on the content of the standard feed - protein and energy amount, antioxidants (vitamin E, C, polyphenolic antioxidants, carotenoids), fatty acids and micronutrient concentrations with similar or opposite effects to the testing additive used in an experiment. In the first study, selenized yeast was added to the diet to identify its effects on gene expression. Selenium is a cofactor of the antioxidant enzymes glutathione peroxidases. Other antioxidants that share a similar mode of action with selenium influence the gene expression in the control group and the alterations in the treatment group turn out different. Thus, in comparison with a medicinal compound, consuming a diet drastically increases the number of endpoints that are capable of influencing the phenotype, due to interactions between ingredients [1].

The use of gene expression analysis provides a very informative tool for the detection of effects caused by a nutritional supplement. The use of microarrays, especially in the beginning of a project, is a very valuable tool to gain an overview about all possible feed supplement effects in an organ or on the cellular level. Based on the gene expression analysis, the appropriate tests on the level of proteins and metabolites can subsequently be performed. This proceeding is very target-oriented and allows the discovery of new biomarkers, changes of protein and metabolite amounts, and altered activity status of transcription factors or signal transducers. Even though the biological interpretation of gene expression analysis is time consuming, due to its complexity, the indications gained can be verified very efficiently, because of the precise knowledge about the involved genes. The author would highly recommend the use of microarrays for animal nutrition research because of its effectiveness, target-oriented strategy, and holistic approach for the discovery of overall effects.

6. References

1. Mutch DM, Wahli W, Williamson G: **Nutrigenomics and nutrigenetics: the emerging faces of nutrition.** *FASEB J* 2005, **19**(12):1602-1616.
2. Dawson KA: **Nutrigenomics: feeding the genes for improved fertility.** *Anim Reprod Sci* 2006, **96**(3-4):312-322.
3. Afman L, Muller M: **Nutrigenomics: from molecular nutrition to prevention of disease.** *J Am Diet Assoc* 2006, **106**(4):569-576.
4. Masotti A, Da Sacco L, Bottazzo GF, Alisi A: **Microarray technology: a promising tool in nutrigenomics.** *Crit Rev Food Sci Nutr* 2010, **50**(7):693-698.
5. [www.mged.org/Workgroups/MIAME/miame.html]
6. Simopoulos AP: **Nutrigenetics/Nutrigenomics.** *Annu Rev Public Health* 2010, **31**:53-68.
7. Roberts RM, Smith GW, Bazer FW, Cibelli J, Seidel GE, Jr., Bauman DE, Reynolds LP, Ireland JJ: **Research priorities. Farm animal research in crisis.** *Science* 2009, **324**(5926):468-469.
8. Suzuki Y, Yeung AC, Ikeno F: **The representative porcine model for human cardiovascular disease.** *J Biomed Biotechnol* 2011, **2011**:195483.
9. Richardson MA, Read LL, Taylor Clelland CL, Reilly MA, Chao HM, Guynn RW, Suckow RF, Clelland JD: **Evidence for a tetrahydrobiopterin deficit in schizophrenia.** *Neuropsychobiology* 2005, **52**(4):190-201.
10. Rayman MP: **Food-chain selenium and human health: emphasis on intake.** *Brit J Nutr* 2008, **100**(2):254-268.
11. Rotter BA, Prelusky DB, Pestka JJ: **Toxicology of deoxynivalenol (vomitoxin).** *Journal of Toxicology and Environmental Health* 1996, **48**(1):1-34.
12. Awad WA, Ghareeb K, Bohm J, Razzazi E, Hellweg P, Zentek J: **The impact of the Fusarium toxin deoxynivalenol (DON) on poultry.** *International Journal of Poultry Science* 2008, **7**(9):827-842.
13. Pestka JJ: **Deoxynivalenol: mechanisms of action, human exposure, and toxicological relevance.** *Arch Toxicol* 2010, **84**(9):663-679.
14. Pappas AC, Zoidis E, Surai PF, Zervas G: **Selenoproteins and maternal nutrition.** *Comparative Biochemistry and Physiology Part B, Biochemistry and Molecular Biology* 2008, **151**(4):361-372.
15. Hoekstra WG: **Biochemical Function of Selenium and Its Relation to Vitamin-E.** *Federation Proceedings* 1975, **34**(11):2083-2089.
16. Steinbrenner H, Sies H: **Protection against reactive oxygen species by selenoproteins.** *Biochim Biophys Acta-General Subjects* 2009, **1790**(11):1478-1485.
17. Jaeschke H: **Mechanisms of oxidant stress-induced acute tissue injury.** *Proceedings of the Society for Experimental Biology and Medicine* 1995, **209**(2):104-111.
18. Rayman MP: **The use of high-selenium yeast to raise selenium status: how does it measure up?** *Brit J Nutr* 2004, **92**(4):557-573.
19. Schafer K, Kyriakopoulos A, Gessner H, Grune T, Behne D: **Effects of selenium deficiency on fatty acid metabolism in rats fed fish oil-enriched diets.** *Journal of Trace Elements in Medicine and Biology* 2004, **18**(1):89-97.
20. Scott TA, Hall JW: **Using acid insoluble ash marker ratios (diet : digesta) to predict digestibility of wheat and barley metabolizable energy and nitrogen retention in broiler chicks.** *Poultry Science* 1998, **77**(5):674-679.
21. Rodriguez EM, Sanz MT, Romero CD: **Critical study of fluorimetric determination of selenium in urine.** *Talanta* 1994, **41**(12):2025-2031.
22. Aeschbacher K, Messikommer R, Meile L, Wenk C: **Bt176 corn in poultry nutrition: physiological characteristics and fate of recombinant plant DNA in chickens.** *Poultry Science* 2005, **84**(3):385-394.
23. **Enzymatic Assay: Glutathione Peroxidase** [www.sigmaaldrich.com/img/assets/18240/Glutathione_Peroxidase.pdf]

24. Zuberbuehler CA, Messikommer RE, Arnold MM, Forrer RS, Wenk C: **Effects of selenium depletion and selenium repletion by choice feeding on selenium status of young and old laying hens.** *Physiology & Behavior* 2006, **87**(2):430-440.
25. Kaiserova K, Srivastava S, Hoetker JD, Awe SO, Tang XL, Cai J, Bhatnagar A: **Redox activation of aldose reductase in the ischemic heart.** *J Biol Chem* 2006, **281**(22):15110-15120.
26. Hara A, Radin NS: **Lipid extraction of tissues with a low-toxicity solvent.** *Analytical Biochemistry* 1978, **90**(1):420-426.
27. Mondello L, Tranchida PQ, Dugo P, Dugo G: **Rapid, micro-scale preparation and very fast gas chromatographic separation of cod liver oil fatty acid methyl esters.** *J Pharm Biomed Anal* 2006, **41**(5):1566-1570.
28. Hoffman DR, Birch EE, Birch DG, Uauy R, Castaneda YS, Lapus MG, Wheaton DH: **Impact of early dietary intake and blood lipid composition of long-chain polyunsaturated fatty acids on later visual development.** *J Pediatr Gastroenterol Nutr* 2000, **31**(5):540-553.
29. Chaudhary M, Garg AK, Mittal GK, Mudgal V: **Effect of organic selenium supplementation on growth, Se uptake, and nutrient utilization in guinea pigs.** *Biological Trace Element Research* 2009, **133**(2):217-226.
30. Holben DH, Smith AM: **The diverse role of selenium within selenoproteins: A review.** *Journal of the American Dietetic Association* 1999, **99**(7):836-843.
31. Fischer A, Pallauf J, Gohil K, Weber SU, Packer L, Rimbach G: **Effect of selenium and vitamin E deficiency on differential gene expression in rat liver.** *Biochem Bioph Res Co* 2001, **285**(2):470-475.
32. Arthur JR: **The glutathione peroxidases.** *Cellular and Molecular Life Sciences* 2000, **57**(13-14):1825-1835.
33. Imai H, Narashima K, Arai M, Sakamoto H, Chiba N, Nakagawa Y: **Suppression of leukotriene formation in RBL-2H3 cells that overexpressed phospholipid hydroperoxide glutathione peroxidase.** *Journal of Biological Chemistry* 1998, **273**(4):1990-1997.
34. Birringer M, Pilawa S, Flohe L: **Trends in selenium biochemistry.** *Nat Prod Rep* 2002, **19**(6):693-718.
35. Pennacchio LA, Olivier M, Hubacek JA, Cohen JC, Cox DR, Fruchart JC, Krauss RM, Rubin EM: **An apolipoprotein influencing triglycerides in humans and mice revealed by comparative sequencing.** *Science* 2001, **294**(5540):169-173.
36. Dichlberger A, Cogburn LA, Nimpf J, Schneider WJ: **Avian apolipoprotein A-V binds to LDL receptor gene family members.** *J Lipid Res* 2007, **48**(7):1451-1456.
37. Yuan J, Palioura S, Salazar JC, Su D, O'Donoghue P, Hohn MJ, Cardoso AM, Whitman WB, Soll D: **RNA-dependent conversion of phosphoserine forms selenocysteine in eukaryotes and archaea.** *Proceedings of the National Academy of Sciences of the United States of America* 2006, **103**(50):18923-18927.
38. Mihara H, Kurihara T, Watanabe T, Yoshimura T, Esaki N: **cDNA cloning, purification, and characterization of mouse liver selenocysteine lyase - Candidate for selenium delivery protein in selenoprotein synthesis.** *Journal of Biological Chemistry* 2000, **275**(9):6195-6200.
39. Combs GF, Combs SB: **The Nutritional Biochemistry of Selenium.** *Annual Review of Nutrition* 1984, **4**:257-280.
40. Bierla K, Dernovics M, Vacchina V, Szpunar J, Bertin G, Lobinski R: **Determination of selenocysteine and selenomethionine in edible animal tissues by 2D size-exclusion reversed-phase HPLC-ICP MS following carbamidomethylation and proteolytic extraction.** *Anal Bioanal Chem* 2008, **390**(7):1789-1798.
41. Pierik M, Rutgeerts P, Vlietinck R, Vermeire S: **Pharmacogenetics in inflammatory bowel disease.** *World J Gastroenterology* 2006, **12**(23):3657-3667.

42. Ranjard L, Nazaret S, Cournoyer B: **Freshwater bacteria can methylate selenium through the thiopurine methyltransferase pathway**. *Applied and Environmental Microbiology* 2003, **69**(7):3784-3790.
43. Cara Terribas CJ, Gonzalez Guijarro L: **[Hypomethylation and multiple sclerosis, the susceptibility factor?]**. *Neurologia* 2002, **17**(3):132-135.
44. Musayev FN, Di Salvo ML, Ko TP, Gandhi AK, Goswami A, Schirch V, Safo MK: **Crystal Structure of human pyridoxal kinase: Structural basis of M+ and M2+ activation**. *Protein Science* 2007, **16**(10):2184-2194.
45. Hackett NR, Heguy A, Harvey BG, O'Connor TP, Luettich K, Flieder DB, Kaplan R, Crystal RG: **Variability of antioxidant-related gene expression in the airway epithelium of cigarette smokers**. *American Journal of Respiratory Cell and Molecular Biology* 2003, **29**(3 Pt 1):331-343.
46. Mahmoud KZ, Edens FW: **Influence of selenium sources on age-related and mild heat stress-related changes of blood and liver glutathione redox cycle in broiler chickens (Gallus domesticus)**. *Comp Biochem Phys B* 2003, **136**(4):921-934.
47. Christensen MJ, Nelson BL, Wray CD: **Regulation of glutathione S-transferase gene expression and activity by dietary selenium**. *Biochem Biophys Res Commun* 1994, **202**(1):271-277.
48. Finley D, Ciechanover A, Varshavsky A: **Ubiquitin as a central cellular regulator**. *Cell* 2004, **116**(2 Suppl):S29-32, 22 p following S32.
49. Dudek EJ, Shang F, Liu Q, Valverde P, Hobbs M, Taylor A: **Selectivity of the ubiquitin pathway for oxidatively modified proteins: relevance to protein precipitation diseases**. *Faseb Journal* 2005, **19**(10):1707-+.
50. Cenciarelli C, Chiaur DS, Guardavaccaro D, Parks W, Vidal M, Pagano M: **Identification of a family of human F-box proteins**. *Current Biology* 1999, **9**(20):1177-1179.
51. Bornhauser BC, Johansson C, Lindholm D: **Functional activities and cellular localization of the ezrin, radixin, moesin (ERM) and RING zinc finger domains in MIR**. *FEBS Letters* 2003, **553**(1-2):195-199.
52. Winston JT, Koepp DM, Zhu C, Elledge SJ, Harper JW: **A family of mammalian F-box proteins**. *Current Biology* 1999, **9**(20):1180-1182.
53. Rajendra R, Malegaonkar D, Pungaliya P, Marshall H, Rasheed Z, Brownell J, Liu LF, Lutzker S, Saleem A, Rubin EH: **Topors functions as an E3 ubiquitin ligase with specific E2 enzymes and ubiquitinates p53**. *Journal of Biological Chemistry* 2004, **279**(35):36440-36444.
54. Cioce M, Boulon S, Matera AG, Lamond AI: **UV-induced fragmentation of Cajal bodies**. *Journal of Cell Biology* 2006, **175**(3):401-413.
55. Hoefer MM, Boneberg EM, Grotegut S, Kusch J, Illges H: **Possible tetramerisation of the proteasome maturation factor POMP/proteassemblin/hUmp1 and its subcellular localisation**. *International Journal of Biological Macromolecules* 2006, **38**(3-5):259-267.
56. Santelli E, Leone M, Li CL, Fukushima T, Preece NE, Olson AJ, Ely KR, Reed JC, Pellecchia M, Liddington RC *et al*: **Structural analysis of Siah1-Siah-interacting protein interactions and insights into the assembly of an E3 ligase multiprotein complex**. *Journal of Biological Chemistry* 2005, **280**(40):34278-34287.
57. Pickart CM: **Back to the future with ubiquitin**. *Cell* 2004, **116**(2):181-190.
58. Yoon I, Werner TM, Butler JM: **Effect of source and concentration of selenium on growth performance and selenium retention in broiler chickens**. *Poult Sci* 2007, **86**(4):727-730.
59. Kenmochi N, Suzuki T, Uechi T, Magoori M, Kuniba M, Higa S, Watanabe K, Tanaka T: **The human mitochondrial ribosomal protein genes: mapping of 54 genes to the chromosomes and implications for human disorders**. *Genomics* 2001, **77**(1-2):65-70.
60. Kannan K, Jain SK: **Oxidative stress and apoptosis**. *Pathophysiology* 2000, **7**(3):153-163.

61. Sauer H, Wartenberg M, Hescheler J: **Reactive oxygen species as intracellular messengers during cell growth and differentiation**. *Cellular Physiology and Biochemistry* 2001, **11**(4):173-186.
62. Silver DL, Wang N, Vogel S: **Identification of small PDZK1-associated protein, DD96/MAP17, as a regulator of PDZK1 and plasma high density lipoprotein levels**. *Journal of Biological Chemistry* 2003, **278**(31):28528-28532.
63. Guijarro MV, Leal JFM, Blanco-Aparicio C, Alonso S, Fominaya J, Lleonart M, Castellvi J, Cajal SRY, Carnero A: **MAP17 enhances the malignant behavior of tumor cells through ROS increase**. *Carcinogenesis* 2007, **28**(10):2096-2104.
64. Wymann MP, Marone R: **Phosphoinositide 3-kinase in disease: timing, location, and scaffolding**. *Current Opinion in Cell Biology* 2005, **17**(2):141-149.
65. Wang JF, Zhang XF, Groopman JE: **Activation of vascular endothelial growth factor receptor-3 and its downstream signaling promote cell survival under oxidative stress**. *Journal of Biological Chemistry* 2004, **279**(26):27088-27097.
66. Matsumoto T, Claesson-Welsh L: **VEGF receptor signal transduction**. *Sci STKE* 2001, **2001**(112):RE21.
67. Pombo CM, Bonventre JV, Molnar A, Kyriakis J, Force T: **Activation of a human Ste20-like kinase by oxidant stress defines a novel stress response pathway**. *EMBO J* 1996, **15**(17):4537-4546.
68. O'Connor T, Ireland LS, Harrison DJ, Hayes JD: **Major differences exist in the function and tissue-specific expression of human aflatoxin B1 aldehyde reductase and the principal human aldo-keto reductase AKR1 family members**. *Biochemical Journal* 1999, **343 Pt 2**:487-504.
69. Jin Y, Penning TM: **Aldo-keto reductases and bioactivation/detoxication**. *Annu Rev Pharmacol Toxicol* 2007, **47**:263-292.
70. Reinartz A, Ehling J, Leue A, Liedtke C, Schneider U, Kopitz J, Weiss T, Hellerbrand C, Weiskirchen R, Knuchel R *et al*: **Lipid-induced up-regulation of human acyl-CoA synthetase 5 promotes hepatocellular apoptosis**. *Biochimica et Biophysica Acta* 2010.
71. Leonard AE, Bobik EG, Dorado J, Kroeger PE, Chuang LT, Thurmond JM, Parker-Barnes JM, Das T, Huang YS, Mukerji P: **Cloning of a human cDNA encoding a novel enzyme involved in the elongation of long-chain polyunsaturated fatty acids**. *Biochemical Journal* 2000, **350 Pt 3**:765-770.
72. Cho HP, Nakamura M, Clarke SD: **Cloning, expression, and fatty acid regulation of the human delta-5 desaturase**. *J Biol Chem* 1999, **274**(52):37335-37339.
73. Korniluk K, Czauderna M, Kowalczyk J, Mieczkowska A, Taciak M, Leng L: **Influence of dietary conjugated linoleic acid isomers and selenium on growth, feed efficiency, and liver fatty acid profile in rats**. *J Anim Feed Sci* 2006, **15**(1):131-146.
74. Raabe M, Veniant MM, Sullivan MA, Zlot CH, Bjorkegren J, Nielsen LB, Wong JS, Hamilton RL, Young SG: **Analysis of the role of microsomal triglyceride transfer protein in the liver of tissue-specific knockout mice**. *J Clin Invest* 1999, **103**(9):1287-1298.
75. Au WS, Kung HF, Lin MC: **Regulation of microsomal triglyceride transfer protein gene by insulin in HepG2 cells: roles of MAPKerk and MAPKp38**. *Diabetes* 2003, **52**(5):1073-1080.
76. Salas-Cortes L, Ye F, Tenza D, Wilheim C, Theos A, Louvard D, Raposo G, Coudrier E: **Myosin 1b modulates the morphology and the protein transport within multi-vesicular sorting endosomes**. *J Cell Sci* 2005, **118**(20):4823-4832.
77. Bock JB, Matern HT, Peden AA, Scheller RH: **A genomic perspective on membrane compartment organization**. *Nature* 2001, **409**(6822):839-841.
78. Kouadio JH, Mobio TA, Baudrimont I, Moukha S, Dano SD, Creppy EE: **Comparative study of cytotoxicity and oxidative stress induced by deoxynivalenol, zearalenone or fumonisin B1 in human intestinal cell line Caco-2**. *Toxicology* 2005, **213**(1-2):56-65.

79. Zou H, Thomas SM, Yan ZW, Grandis JR, Vogt A, Li LY: **Human rhomboid family-1 gene RHBDF1 participates in GPCR-mediated transactivation of EGFR growth signals in head and neck squamous cancer cells.** *FASEB J* 2009, **23**(2):425-432.
80. Meyer HA, Grau H, Kraft R, Kostka S, Prehn S, Kalies KU, Hartmann E: **Mammalian Sec61 is associated with Sec62 and Sec63.** *J Biol Chem* 2000, **275**(19):14550-14557.
81. Nell S, Bahtz R, Bossecker A, Kipp A, Landes N, Bumke-Vogt C, Halligan E, Lunec J, Brigelius-Flohe R: **PCR-verified microarray analysis and functional in vitro studies indicate a role of alpha-tocopherol in vesicular transport.** *Free Radical Research* 2007, **41**(8):930-942.
82. Dunn AY, Melville MW, Frydman J: **Review: cellular substrates of the eukaryotic chaperonin TRiC/CCT.** *J Struct Biol* 2001, **135**(2):176-184.
83. Helenius A, Aebi M: **Roles of N-linked glycans in the endoplasmic reticulum.** *Annual Review of Biochemistry* 2004, **73**:1019-1049.
84. Meusser B, Hirsch C, Jarosch E, Sommer T: **ERAD: the long road to destruction.** *Nat Cell Biol* 2005, **7**(8):766-772.
85. Katiyar S, Li GT, Lennarz WJ: **A complex between peptide : N-glycanase and two proteasome-linked proteins suggests a mechanism for the degradation of misfolded glycoproteins.** *Proceedings of the National Academy of Sciences of the United States of America* 2004, **101**(38):13774-13779.
86. Dietrich B, Neuenschwander S, Bucher B, Wenk C: **Fusarium mycotoxin-contaminated wheat containing deoxynivalenol alters the gene expression in the liver and the jejunum of broilers.** *animal* 2011, **FirstView**:1-14.
87. Awad WA, Bohm J, Razzazi-Fazeli E, Hulan HW, Zentek J: **Effects of deoxynivalenol on general performance and electrophysiological properties of intestinal mucosa of broiler chickens.** *Poultry Science* 2004, **83**(12):1964-1972.
88. European Union: **2006/576/EC: Commission Recommendation of 17 August 2006 on the presence of deoxynivalenol, zearalenone, ochratoxin A, T-2 and HT-2 and fumonisins in products intended for animal feeding.** *Official Journal - European Union Legislation* 2006, **49**:7-9.
89. Swamy HV, Smith TK, Karrow NA, Boermans HJ: **Effects of feeding blends of grains naturally contaminated with Fusarium mycotoxins on growth and immunological parameters of broiler chickens.** *Poultry Science* 2004, **83**(4):533-543.
90. Prelusky DB, Gerdes RG, Underhill KL, Rotter BA, Jui PY, Trenholm HL: **Effects of low-level dietary deoxynivalenol on haematological and clinical parameters of the pig.** *Natural Toxins* 1994, **2**(3):97-104.
91. Lun AK, Moran ET, Jr., Young LG, McMillan EG: **Absorption and elimination of an oral dose of 3H-deoxynivalenol in colostomized and intact chickens.** *Bulletin of Environmental Contamination and Toxicology* 1989, **42**(6):919-925.
92. Yoshizawa T, Cote LM, Swanson SP, and Buck WB: **Confirmation of DOM-1, a de-epoxidation metabolite of deoxynivalenol, in biological fluids of lactating cows.** *Agricultural and Biological Chemistry* 1986, **50**(1):227-229.
93. Pestka JJ: **Deoxynivalenol: Toxicity, mechanisms and animal health risks.** *Animal Feed Science and Technology* 2007, **137**(3-4):283-298.
94. Ueno Y: **Toxicological features of T-2 toxin and related trichothecenes.** *Fundamental and Applied Toxicology* 1984, **4**(2 Pt 2):S124-132.
95. Bondy GS, Pestka JJ: **Immunomodulation by fungal toxins.** *Journal of Toxicology and Environmental Health Part B, Critical Reviews* 2000, **3**(2):109-143.
96. National ResearchCouncil (ed.): **Nutrient Requirements of Poultry.** Washington, DC: National Academy Press; 1994.
97. Pepper SD, Saunders EK, Edwards LE, Wilson CL, Miller CJ: **The utility of MAS5 expression summary and detection call algorithms.** *BMC Bioinformatics* 2007, **8**:273.

98. EFSA: **Opinion of the Scientific Panel on Contaminants in the Food Chain on a request from the Commission related to Deoxynivalenol as undesirable substance in animal feed**. *The EFSA journal* 2004, **89**:1-35.

99. Maresca M, Mahfoud R, Garmy N, Fantini J: **The mycotoxin deoxynivalenol affects nutrient absorption in human intestinal epithelial cells**. *Journal of Nutrition* 2002, **132**(9):2723-2731.

100. Stahl A, Hirsch DJ, Gimeno RE, Punreddy S, Ge P, Watson N, Patel S, Kotler M, Raimondi A, Tartaglia LA *et al*: **Identification of the major intestinal fatty acid transport protein**. *Molecular Cell* 1999, **4**(3):299-308.

101. Nakauchi J, Matsuo H, Kim DK, Goto A, Chairoungdua A, Cha SH, Inatomi J, Shiokawa Y, Yamaguchi K, Saito I *et al*: **Cloning and characterization of a human brain Na(+)-independent transporter for small neutral amino acids that transports D-serine with high affinity**. *Neuroscience Letters* 2000, **287**(3):231-235.

102. Kanai Y, Segawa H, Miyamoto K, Uchino H, Takeda E, Endou H: **Expression cloning and characterization of a transporter for large neutral amino acids activated by the heavy chain of 4F2 antigen (CD98)**. *Journal of Biological Chemistry* 1998, **273**(37):23629-23632.

103. Robbana-Barnat S, Loridon-Rosa B, Cohen H, Lafarge-Frayssinet C, Neish GA, Frayssinet C: **Protein synthesis inhibition and cardiac lesions associated with deoxynivalenol ingestion in mice**. *Food Additives and Contaminants* 1987, **4**(1).

104. De Walle JV, Sergent T, Piront N, Toussaint O, Schneider YJ, Larondelle Y: **Deoxynivalenol affects in vitro intestinal epithelial cell barrier integrity through inhibition of protein synthesis**. *Toxicol Appl Pharmacol* 2010, **245**(3):291-298.

105. Lambert DW, Wood IS, Ellis A, Shirazi-Beechey SP: **Molecular changes in the expression of human colonic nutrient transporters during the transition from normality to malignancy**. *British Journal of Cancer* 2002, **86**(8):1262-1269.

106. Jin Y, Penning TM: **Aldo-keto reductases and bioactivation/detoxication**. *Annual Review of Pharmacology and Toxicology* 2007, **47**:263-292.

107. Kawata K, Yokoo H, Shimazaki R, Okabe S: **Classification of heavy-metal toxicity by human DNA microarray analysis**. *Environmental Science and Technology* 2007, **41**(10):3769-3774.

108. Plummer ER, Calvert H: **Targeting poly(ADP-ribose) polymerase: a two-armed strategy for cancer therapy**. *Clinical Cancer Research* 2007, **13**(21):6252-6256.

109. Shifrin VI, Anderson P: **Trichothecene mycotoxins trigger a ribotoxic stress response that activates c-Jun N-terminal kinase and p38 mitogen-activated protein kinase and induces apoptosis**. *Journal of Biological Chemistry* 1999, **274**(20):13985-13992.

110. Marzocco S, Russo R, Bianco G, Autore G, Severino L: **Pro-apoptotic effects of nivalenol and deoxynivalenol trichothecenes in J774A.1 murine macrophages**. *Toxicology Letters* 2009, **189**(1):21-26.

111. O'Brien PJ, Ellenberger T: **Dissecting the broad substrate specificity of human 3-methyladenine-DNA glycosylase**. *Journal of Biological Chemistry* 2004, **279**(11):9750-9757.

112. Abraham J, Lemmers B, Hande MP, Moynahan ME, Chahwan C, Ciccia A, Essers J, Hanada K, Chahwan R, Khaw AK *et al*: **Eme1 is involved in DNA damage processing and maintenance of genomic stability in mammalian cells**. *EMBO Journal* 2003, **22**(22):6137-6147.

113. Miller KA, Yoshikawa DM, McConnell IR, Clark R, Schild D, Albala JS: **RAD51C interacts with RAD51B and is central to a larger protein complex in vivo exclusive of RAD51**. *Journal of Biological Chemistry* 2002, **277**(10):8406-8411.

114. Mello JA, Sillje HH, Roche DM, Kirschner DB, Nigg EA, Almouzni G: **Human Asf1 and CAF-1 interact and synergize in a repair-coupled nucleosome assembly pathway**. *EMBO Reports* 2002, **3**(4):329-334.

115. Silverman J, Takai H, Buonomo SB, Eisenhaber F, de Lange T: **Human Rif1, ortholog of a yeast telomeric protein, is regulated by ATM and 53BP1 and functions in the S-phase checkpoint.** *Genes and Development* 2004, **18**(17):2108-2119.
116. Shimamoto T, Tanimura T, Yoneda Y, Kobayakawa Y, Sugasawa K, Hanaoka F, Oka M, Okada Y, Tanaka K, Kohno K: **Expression and functional analyses of the Dxpa gene, the Drosophila homolog of the human excision repair gene XPA.** *Journal of Biological Chemistry* 1995, **270**(38):22452-22459.
117. Frankic T, Pajk T, Rezar V, Levart A, Salobir J: **The role of dietary nucleotides in reduction of DNA damage induced by T-2 toxin and deoxynivalenol in chicken leukocytes.** *Food and Chemical Toxicology* 2006, **44**(11):1838-1844.
118. Yang GH, Li S, Pestka JJ: **Down-regulation of the endoplasmic reticulum chaperone GRP78/BiP by vomitoxin (Deoxynivalenol).** *Toxicology and Applied Pharmacology* 2000, **162**(3):207-217.
119. Ron D: **Translational control in the endoplasmic reticulum stress response.** *Journal of Clinical Investigation* 2002, **110**(10):1383-1388.
120. van Huizen R, Martindale JL, Gorospe M, Holbrook NJ: **P58IPK, a novel endoplasmic reticulum stress-inducible protein and potential negative regulator of eIF2alpha signaling.** *Journal of Biological Chemistry* 2003, **278**(18):15558-15564.
121. Gale M, Jr., Blakely CM, Hopkins DA, Melville MW, Wambach M, Romano PR, Katze MG: **Regulation of interferon-induced protein kinase PKR: modulation of P58IPK inhibitory function by a novel protein, P52rIPK.** *Molecular and Cellular Biology* 1998, **18**(2):859-871.
122. Pestka JJ: **Deoxynivalenol: mechanisms of action, human exposure, and toxicological relevance.** *Archives of Toxicology* 2010, **84**(9):663-679.
123. Mukherjee D, Gao M, O'Connor JP, Raijmakers R, Pruijn G, Lutz CS, Wilusz J: **The mammalian exosome mediates the efficient degradation of mRNAs that contain AU-rich elements.** *EMBO Journal* 2002, **21**(1-2):165-174.
124. van Dijk EL, Schilders G, Pruijn GJ: **Human cell growth requires a functional cytoplasmic exosome, which is involved in various mRNA decay pathways.** *RNA* 2007, **13**(7):1027-1035.
125. Schilders G, van Dijk E, Raijmakers R, Pruijn GJ: **Cell and molecular biology of the exosome: how to make or break an RNA.** *International Review of Cytology* 2006, **251**:159-208.
126. Pestka JJ: **Mechanisms of deoxynivalenol-induced gene expression and apoptosis.** *Food Additives and Contaminants* 2008:1-13.
127. Chen CY, Gherzi R, Ong SE, Chan EL, Raijmakers R, Pruijn GJ, Stoecklin G, Moroni C, Mann M, Karin M: **AU binding proteins recruit the exosome to degrade ARE-containing mRNAs.** *Cell* 2001, **107**(4):451-464.
128. Paschoud S, Dogar AM, Kuntz C, Grisoni-Neupert B, Richman L, Kuhn LC: **Destabilization of interleukin-6 mRNA requires a putative RNA stem-loop structure, an AU-rich element, and the RNA-binding protein AUF1.** *Molecular and Cellular Biology* 2006, **26**(22):8228-8241.
129. Rajasingh J, Bord E, Luedemann C, Asai J, Hamada H, Thorne T, Qin G, Goukassian D, Zhu Y, Losordo DW *et al*: **IL-10-induced TNF-alpha mRNA destabilization is mediated via IL-10 suppression of p38 MAP kinase activation and inhibition of HuR expression.** *FASEB Journal* 2006, **20**(12):2112-2114.
130. Sergent T, Parys M, Garsou S, Pussemier L, Schneider YJ, Larondelle Y: **Deoxynivalenol transport across human intestinal Caco-2 cells and its effects on cellular metabolism at realistic intestinal concentrations.** *Toxicology Letters* 2006, **164**(2):167-176.
131. Diesing AK, Nossol C, Panther P, Walk N, Post A, Kluess J, Kreutzmann P, Danicke S, Rothkotter HJ, Kahlert S: **Mycotoxin deoxynivalenol (DON) mediates biphasic cellular**

response in intestinal porcine epithelial cell lines IPEC-1 and IPEC-J2. *Toxicology Letters* 2011, **200**(1-2):8-18.
132. Harhaj NS, Antonetti DA: **Regulation of tight junctions and loss of barrier function in pathophysiology**. *International Journal of Biochemistry & Cell Biology* 2004, **36**(7):1206-1237.
133. Pinton P, Nougayrede JP, Del Rio JC, Moreno C, Marin DE, Ferrier L, Bracarense AP, Kolf-Clauw M, Oswald IP: **The food contaminant deoxynivalenol, decreases intestinal barrier permeability and reduces claudin expression**. *Toxicol Appl Pharmacol* 2009, **237**(1):41-48.
134. Bosserhoff AK, Moser M, Scholmerich J, Buettner R, Hellerbrand C: **Specific expression and regulation of the new melanoma inhibitory activity-related gene MIA2 in hepatocytes**. *Journal of Biological Chemistry* 2003, **278**(17):15225-15231.
135. Takezako N, Hayakawa M, Hayakawa H, Aoki S, Yanagisawa K, Endo H, Tominaga S: **ST2 suppresses IL-6 production via the inhibition of IkappaB degradation induced by the LPS signal in THP-1 cells**. *Biochemical and Biophysical Research Communications* 2006, **341**(2):425-432.
136. Wong S, Schwartz RC, Pestka JJ: **Superinduction of TNF-alpha and IL-6 in macrophages by vomitoxin (deoxynivalenol) modulated by mRNA stabilization**. *Toxicology* 2001, **161**(1-2):139-149.
137. Swamy HV, Smith TK, Cotter PF, Boermans HJ, Sefton AE: **Effects of feeding blends of grains naturally contaminated with Fusarium mycotoxins on production and metabolism in broilers**. *Poultry Science* 2002, **81**(7):966-975.
138. Azcona-Olivera JI, Ouyang Y, Murtha J, Chu FS, Pestka JJ: **Induction of cytokine mRNAs in mice after oral exposure to the trichothecene vomitoxin (deoxynivalenol): relationship to toxin distribution and protein synthesis inhibition**. *Toxicology and Applied Pharmacology* 1995, **133**(1):109-120.
139. Liu X, Xiong F, Wei X, Yang H, Zhou R: **LAPTM4B-35, a novel tetratransmembrane protein and its PPRP motif play critical roles in proliferation and metastatic potential of hepatocellular carcinoma cells**. *Cancer* 2009, **100**(12):2335-2340.
140. Xu CS, Chang CF, Yuan JY, Li WQ, Han HP, Yang KJ, Zhao LF, Li YC, Zhang HY, Rahman S *et al*: **Expressed genes in regenerating rat liver after partial hepatectomy**. *World Journal of Gastroenterology* 2005, **11**(19):2932-2940.
141. Ricci JE, Munoz-Pinedo C, Fitzgerald P, Bailly-Maitre B, Perkins GA, Yadava N, Scheffler IE, Ellisman MH, Green DR: **Disruption of mitochondrial function during apoptosis is mediated by caspase cleavage of the p75 subunit of complex I of the electron transport chain**. *Cell* 2004, **117**(6):773-786.
142. Loeffen JL, Triepels RH, van den Heuvel LP, Schuelke M, Buskens CA, Smeets RJ, Trijbels JM, Smeitink JA: **cDNA of eight nuclear encoded subunits of NADH:ubiquinone oxidoreductase: human complex I cDNA characterization completed**. *Biochem Biophys Res Commun* 1998, **253**(2):415-422.
143. Wang Z, Spittau B, Behrendt M, Peters B, Krieglstein K: **Human TIEG2/KLF11 induces oligodendroglial cell death by downregulation of Bcl-X(L) expression**. *J Neural Transm* 2007, **114**(7):867-875.
144. Chung YJ, Zhou HR, Pestka JJ: **Transcriptional and posttranscriptional roles for p38 mitogen-activated protein kinase in upregulation of TNF-alpha expression by deoxynivalenol (vomitoxin)**. *Toxicol Appl Pharmacol* 2003, **193**(2):188-201.
145. Pan W, Zhang Q, Xi QS, Gan RB, Li TP: **FUP1, a gene associated with hepatocellular carcinoma, stimulates NIH3T3 cell proliferation and tumor formation in nude mice**. *Biochem Biophys Res Commun* 2001, **286**(5):1033-1038.
146. Faixova Z, Faix S, Borutova R, Leng L: **Effect of different doses of deoxynivalenol on metabolism in broiler chickens**. *BULLETIN- VETERINARY INSTITUTE IN PULAWY* 2007, **51**(3):421-424.

147. de Wolf CJ, Cupers RM, Bertina RM, Vos HL: **Interleukin-6 induction of protein s is regulated through signal transducer and activator of transcription 3**. *Arterioscler Thromb Vasc Biol* 2006, **26**(9):2168-2174.

148. Huang J, Zhao YL, Li Y, Fletcher JA, Xiao S: **Genomic and functional evidence for an ARID1A tumor suppressor role**. *Genes Chromosomes Cancer* 2007, **46**(8):745-750.

149. Nagl NG, Jr., Wang X, Patsialou A, Van Scoy M, Moran E: **Distinct mammalian SWI/SNF chromatin remodeling complexes with opposing roles in cell-cycle control**. *EMBO J* 2007, **26**(3):752-763.

150. Zhou HR, Islam Z, Pestka JJ: **Induction of competing apoptotic and survival signaling pathways in the macrophage by the ribotoxic trichothecene deoxynivalenol**. *Toxicol Sci* 2005, **87**(1):113-122.

151. Yoon KA, Nakamura Y, Arakawa H: **Identification of ALDH4 as a p53-inducible gene and its protective role in cellular stresses**. *J Hum Genet* 2004, **49**(3):134-140.

152. Scheffner M, Huibregtse JM, Howley PM: **Identification of a human ubiquitin-conjugating enzyme that mediates the E6-AP-dependent ubiquitination of p53**. *Proc Natl Acad Sci U S A* 1994, **91**(19):8797-8801.

153. Nagashima M, Shiseki M, Pedeux RM, Okamura S, Kitahama-Shiseki M, Miura K, Yokota J, Harris CC: **A novel PHD-finger motif protein, p47ING3, modulates p53-mediated transcription, cell cycle control, and apoptosis**. *Oncogene* 2003, **22**(3):343-350.

154. Kawata K, Yokoo H, Shimazaki R, Okabe S: **Classification of heavy-metal toxicity by human DNA microarray analysis**. *Environ Sci Technol* 2007, **41**(10):3769-3774.

155. Stein S, Thomas EK, Herzog B, Westfall MD, Rocheleau JV, Jackson RS, 2nd, Wang M, Liang P: **NDRG1 is necessary for p53-dependent apoptosis**. *J Biol Chem* 2004, **279**(47):48930-48940.

156. Passer BJ, Nancy-Portebois V, Amzallag N, Prieur S, Cans C, Roborel de Climens A, Fiucci G, Bouvard V, Tuynder M, Susini L *et al*: **The p53-inducible TSAP6 gene product regulates apoptosis and the cell cycle and interacts with Nix and the Myt1 kinase**. *Proc Natl Acad Sci U S A* 2003, **100**(5):2284-2289.

157. Shifrin VI, Anderson P: **Trichothecene mycotoxins trigger a ribotoxic stress response that activates c-Jun N-terminal kinase and p38 mitogen-activated protein kinase and induces apoptosis**. *J Biol Chem* 1999, **274**(20):13985-13992.

158. Scott RE, White-Grindley E, Ruley HE, Chesler EJ, Williams RW: **P2P-R expression is genetically coregulated with components of the translation machinery and with PUM2, a translational repressor that associates with the P2P-R mRNA**. *J Cell Physiol* 2005, **204**(1):99-105.

159. Ueno Y: **Toxicological features of T-2 toxin and related trichothecenes**. *Fundam Appl Toxicol* 1984, **4**(2 Pt 2):S124-132.

160. Tanis KQ, Duman RS, Newton SS: **CREB binding and activity in brain: regional specificity and induction by electroconvulsive seizure**. *Biol Psychiatry* 2008, **63**(7):710-720.

161. Kabuyama Y, Oshima K, Kitamura T, Homma M, Yamaki J, Munakata M, Homma Y: **Involvement of selenoprotein P in the regulation of redox balance and myofibroblast viability in idiopathic pulmonary fibrosis**. *Genes Cells* 2007, **12**(11):1235-1244.

162. Rao L, Puschner B, Prolla TA: **Gene expression profiling of low selenium status in the mouse intestine: Transcriptional activation of genes linked to DNA damage, cell cycle control and oxidative stress**. *Journal of Nutrition* 2001, **131**(12):3175-3181.

163. Rotter BA, Prelusky DB, Pestka JJ: **Toxicology of deoxynivalenol (vomitoxin)**. *J Toxicol Environ Health* 1996, **48**(1):1-34.

164. Pestka JJ: **Mechanisms of deoxynivalenol-induced gene expression and apoptosis**. *Food Addit Contam Part A Chem Anal Control Expo Risk Assess* 2008, **25**(9):1128-1140.

165. Prelusky DB, Gerdes RG, Underhill KL, Rotter BA, Jui PY, Trenholm HL: **Effects of low-level dietary deoxynivalenol on haematological and clinical parameters of the pig.** *Nat Toxins* 1994, **2**(3):97-104.
166. Swamy HV, Smith TK, MacDonald EJ: **Effects of feeding blends of grains naturally contaminated with Fusarium mycotoxins on brain regional neurochemistry of starter pigs and broiler chickens.** *J Anim Sci* 2004, **82**(7):2131-2139.
167. Li M, Pestka JJ: **Comparative induction of 28S ribosomal RNA cleavage by ricin and the trichothecenes deoxynivalenol and T-2 toxin in the macrophage.** *Toxicol Sci* 2008, **105**(1):67-78.
168. Diesing AK, Nossol C, Panther P, Walk N, Post A, Kluess J, Kreutzmann P, Danicke S, Rothkotter HJ, Kahlert S: **Mycotoxin deoxynivalenol (DON) mediates biphasic cellular response in intestinal porcine epithelial cell lines IPEC-1 and IPEC-J2.** *Toxicol Lett* 2011, **200**(1-2):8-18.
169. Yang H, Chung DH, Kim YB, Choi YH, Moon Y: **Ribotoxic mycotoxin deoxynivalenol induces G2/M cell cycle arrest via p21Cip/WAF1 mRNA stabilization in human epithelial cells.** *Toxicology* 2008, **243**(1-2):145-154.
170. Gouze ME, Laffitte J, Rouimi P, Loiseau N, Oswald IP, Galtier P: **Effect of various doses of deoxynivalenol on liver xenobiotic metabolizing enzymes in mice.** *Food Chem Toxicol* 2006, **44**(4):476-483.
171. Swamy HV, Smith TK, Karrow NA, Boermans HJ: **Effects of feeding blends of grains naturally contaminated with Fusarium mycotoxins on growth and immunological parameters of broiler chickens.** *Poult Sci* 2004, **83**(4):533-543.
172. Swamy HV, Smith TK, MacDonald EJ, Boermans HJ, Squires EJ: **Effects of feeding a blend of grains naturally contaminated with Fusarium mycotoxins on swine performance, brain regional neurochemistry, and serum chemistry and the efficacy of a polymeric glucomannan mycotoxin adsorbent.** *J Anim Sci* 2002, **80**(12):3257-3267.
173. Yegani M, Smith TK, Leeson S, Boermans HJ: **Effects of feeding grains naturally contaminated with Fusarium mycotoxins on performance and metabolism of broiler breeders.** *Poult Sci* 2006, **85**(9):1541-1549.
174. Scott TA, Hall JW: **Using acid insoluble ash marker ratios (diet:digesta) to predict digestibility of wheat and barley metabolizable energy and nitrogen retention in broiler chicks.** *Poult Sci* 1998, **77**(5):674-679.
175. Irizarry RA, Bolstad BM, Collin F, Cope LM, Hobbs B, Speed TP: **Summaries of Affymetrix GeneChip probe level data.** *Nucleic Acids Res* 2003, **31**(4):e15.
176. Dang H, Liu Y, Pang W, Li C, Wang N, Shyy JY, Zhu Y: **Suppression of 2,3-oxidosqualene cyclase by high fat diet contributes to liver X receptor-alpha-mediated improvement of hepatic lipid profile.** *J Biol Chem* 2009, **284**(10):6218-6226.
177. Awad WA, Bohm J, Razzazi-Fazeli E, Hulan HW, Zentek J: **Effects of deoxynivalenol on general performance and electrophysiological properties of intestinal mucosa of broiler chickens.** *Poult Sci* 2004, **83**(12):1964-1972.
178. Ozkan S, Malayoglu HB, Yalcin S, Karadas F, Kocturk S, Cabuk M, Oktay G, Ozdemir S, Ozdemir E, Ergul M: **Dietary vitamin E (alpha-tocopherol acetate) and selenium supplementation from different sources: performance, ascites-related variables and antioxidant status in broilers reared at low and optimum temperatures.** *Br Poult Sci* 2007, **48**(5):580-593.
179. Puvadolpirod S, Thaxton JP: **Model of physiological stress in chickens 1. Response parameters.** *Poult Sci* 2000, **79**(3):363-369.
180. Pestka JJ, Dong W, Warner RL, Rasooly L, Bondy GS, Brooks KH: **Elevated membrane IgA+ and CD4+ (T helper) populations in murine Peyer's patch and splenic lymphocytes during dietary administration of the trichothecene vomitoxin (deoxynivalenol).** *Food Chem Toxicol* 1990, **28**(6):409-420.

181. Lee KM, Kang D, Clapper ML, Ingelman-Sundberg M, Ono-Kihara M, Kiyohara C, Min S, Lan Q, Le Marchand L, Lin P *et al*: **CYP1A1, GSTM1, and GSTT1 polymorphisms, smoking, and lung cancer risk in a pooled analysis among Asian populations.** *Cancer Epidemiol Biomarkers Prev* 2008, **17**(5):1120-1126.

182. Elizondo G, Fernandez-Salguero P, Sheikh MS, Kim GY, Fornace AJ, Lee KS, Gonzalez FJ: **Altered cell cycle control at the G(2)/M phases in aryl hydrocarbon receptor-null embryo fibroblast.** *Mol Pharmacol* 2000, **57**(5):1056-1063.

183. Gannon M, Gilday D, Rifkind AB: **TCDD induces CYP1A4 and CYP1A5 in chick liver and kidney and only CYP1A4, an enzyme lacking arachidonic acid epoxygenase activity, in myocardium and vascular endothelium.** *Toxicol Appl Pharmacol* 2000, **164**(1):24-37.

184. Shimizu C, Fuda H, Lee YC, Strott CA: **Transcriptional regulation of human 3'-phosphoadenosine 5'-phosphosulphate synthase 2.** *Biochem J* 2002, **363**(Pt 2):263-271.

185. Mastyugin V, Aversa E, Bonazzi A, Vafaes C, Mieyal P, Schwartzman ML: **Hypoxia-induced production of 12-hydroxyeicosanoids in the corneal epithelium: involvement of a cytochrome P-4504B1 isoform.** *J Pharmacol Exp Ther* 1999, **289**(3):1611-1619.

186. Pestka JJ: **Deoxynivalenol: Toxicity, mechanisms and animal health risks.** *Animal feed science and technology* 2007, **137**(3):283.

187. Kinser S, Jia Q, Li M, Laughter A, Cornwell P, Corton JC, Pestka J: **Gene expression profiling in spleens of deoxynivalenol-exposed mice: immediate early genes as primary targets.** *J Toxicol Environ Health A* 2004, **67**(18):1423-1441.

188. Bodea GOD, Munteanu MC, Dinu D, Serban AI, Roming FI, Costache M, Dinischiotu A: **Influence of deoxynivalenol on the oxidative status of HepG2 cells.** *ROMANIAN BIOTECHNOLOGICAL LETTERS* 2009, **14**(2):4349-4359.

189. Ren B, Cam H, Takahashi Y, Volkert T, Terragni J, Young RA, Dynlacht BD: **E2F integrates cell cycle progression with DNA repair, replication, and G(2)/M checkpoints.** *Genes Dev* 2002, **16**(2):245-256.

190. Giacinti C, Giordano A: **RB and cell cycle progression.** *Oncogene* 2006, **25**(38):5220-5227.

191. Santamaria D, Barriere C, Cerqueira A, Hunt S, Tardy C, Newton K, Caceres JF, Dubus P, Malumbres M, Barbacid M: **Cdk1 is sufficient to drive the mammalian cell cycle.** *Nature* 2007, **448**(7155):811-815.

192. Bashir T, Pagano M: **Cdk1: the dominant sibling of Cdk2.** *Nat Cell Biol* 2005, **7**(8):779-781.

193. Yoon CH, Miah MA, Kim KP, Bae YS: **New Cdc2 Tyr 4 phosphorylation by dsRNA-activated protein kinase triggers Cdc2 polyubiquitination and G2 arrest under genotoxic stresses.** *EMBO Rep* 2010, **11**(5):393-399.

194. Pestka JJ, Zhou HR, Moon Y, Chung YJ: **Cellular and molecular mechanisms for immune modulation by deoxynivalenol and other trichothecenes: unraveling a paradox.** *Toxicol Lett* 2004, **153**(1):61-73.

195. Li CL, Lu CY, Ke PY, Chang ZF: **Perturbation of ATP-induced tetramerization of human cytosolic thymidine kinase by substitution of serine-13 with aspartic acid at the mitotic phosphorylation site.** *Biochem Biophys Res Commun* 2004, **313**(3):587-593.

196. Blethrow JD, Glavy JS, Morgan DO, Shokat KM: **Covalent capture of kinase-specific phosphopeptides reveals Cdk1-cyclin B substrates.** *Proc Natl Acad Sci U S A* 2008, **105**(5):1442-1447.

197. Kim EH, Kim SU, Choi KS: **Rottlerin sensitizes glioma cells to TRAIL-induced apoptosis by inhibition of Cdc2 and the subsequent downregulation of survivin and XIAP.** *Oncogene* 2005, **24**(5):838-849.

198. Bibby AC, Litchfield DW: **The multiple personalities of the regulatory subunit of protein kinase CK2: CK2 dependent and CK2 independent roles reveal a secret identity for CK2beta.** *Int J Biol Sci* 2005, **1**(2):67-79.

199. Kreutzer J, Guerra B: **The regulatory beta-subunit of protein kinase CK2 accelerates the degradation of CDC25A phosphatase through the checkpoint kinase Chk1.** *Int J Oncol* 2007, **31**(5):1251-1259.
200. Lin X, Liu S, Luo X, Ma X, Guo L, Li L, Li Z, Tao Y, Cao Y: **EBV-encoded LMP1 regulates Op18/stathmin signaling pathway by cdc2 mediation in nasopharyngeal carcinoma cells.** *Int J Cancer* 2009, **124**(5):1020-1027.
201. Schatz CA, Santarella R, Hoenger A, Karsenti E, Mattaj IW, Gruss OJ, Carazo-Salas RE: **Importin alpha-regulated nucleation of microtubules by TPX2.** *EMBO J* 2003, **22**(9):2060-2070.
202. Gaudet S, Branton D, Lue RA: **Characterization of PDZ-binding kinase, a mitotic kinase.** *Proc Natl Acad Sci U S A* 2000, **97**(10):5167-5172.
203. Dong Z, Liu LH, Han B, Pincheira R, Zhang JT: **Role of eIF3 p170 in controlling synthesis of ribonucleotide reductase M2 and cell growth.** *Oncogene* 2004, **23**(21):3790-3801.
204. Marson A, Kretschmer K, Frampton GM, Jacobsen ES, Polansky JK, MacIsaac KD, Levine SS, Fraenkel E, von Boehmer H, Young RA: **Foxp3 occupancy and regulation of key target genes during T-cell stimulation.** *Nature* 2007, **445**(7130):931-935.
205. Tio L, Villarreal L, Atrian S, Capdevila M: **Functional differentiation in the mammalian metallothionein gene family: metal binding features of mouse MT4 and comparison with its paralog MT1.** *J Biol Chem* 2004, **279**(23):24403-24413.
206. Haq F, Mahoney M, Koropatnick J: **Signaling events for metallothionein induction.** *Mutat Res* 2003, **533**(1-2):211-226.
207. Lim D, Jocelyn KM, Yip GW, Bay BH: **Silencing the Metallothionein-2A gene inhibits cell cycle progression from G1- to S-phase involving ATM and cdc25A signaling in breast cancer cells.** *Cancer Lett* 2009, **276**(1):109-117.
208. Hudson DF, Ohta S, Freisinger T, Macisaac F, Sennels L, Alves F, Lai F, Kerr A, Rappsilber J, Earnshaw WC: **Molecular and genetic analysis of condensin function in vertebrate cells.** *Mol Biol Cell* 2008, **19**(7):3070-3079.
209. Maiorano D, Lutzmann M, Mechali M: **MCM proteins and DNA replication.** *Current Opinion in Cell Biology* 2006, **18**(2):130-136.
210. Yeager M, Harris AL: **Gap junction channel structure in the early 21st century: facts and fantasies.** *Curr Opin Cell Biol* 2007, **19**(5):521-528.
211. Jelen H, Latus-Zietkiewicz D, Wosowicz E, Kaminski E: **Trichodiene as a volatile marker for trichothecenes biosynthesis.** *JOURNAL OF MICROBIOLOGICAL METHODS* 1997, **31**(1/2):45-49.
212. Fujita H, Sugimoto K, Inatomi S, Maeda T, Osanai M, Uchiyama Y, Yamamoto Y, Wada T, Kojima T, Yokozaki H *et al*: **Tight junction proteins claudin-2 and -12 are critical for vitamin D-dependent Ca2+ absorption between enterocytes.** *Mol Biol Cell* 2008, **19**(5):1912-1921.
213. Kansaku A, Hirabayashi S, Mori H, Fujiwara N, Kawata A, Ikeda M, Rokukawa C, Kurihara H, Hata Y: **Ligand-of-Numb protein X is an endocytic scaffold for junctional adhesion molecule 4.** *Oncogene* 2006, **25**(37):5071-5084.
214. Schwartz MW, Woods SC, Porte D, Jr., Seeley RJ, Baskin DG: **Central nervous system control of food intake.** *Nature* 2000, **404**(6778):661-671.
215. Islam Z, Pestka JJ: **Role of IL-1(beta) in endotoxin potentiation of deoxynivalenol-induced corticosterone response and leukocyte apoptosis in mice.** *Toxicol Sci* 2003, **74**(1):93-102.
216. Rodrigues AR, Pignatelli D, Almeida H, Gouveia AM: **Melanocortin 5 receptor activates ERK1/2 through a PI3K-regulated signaling mechanism.** *Mol Cell Endocrinol* 2009, **303**(1-2):74-81.

217. Moon Y, Pestka JJ: **Vomitoxin-induced cyclooxygenase-2 gene expression in macrophages mediated by activation of ERK and p38 but not JNK mitogen-activated protein kinases.** *Toxicological Sciences* 2002, **69**(2):373-382.

218. Wu CS, Greenwood DR, Cooney JM, Jensen DJ, Tatnell MA, Cooper GJ, Mountjoy KG: **Peripherally administered desacetyl alpha-MSH and alpha-MSH both influence postnatal rat growth and associated rat hypothalamic protein expression.** *Am J Physiol Endocrinol Metab* 2006, **291**(6):E1372-1380.

219. Pons R, Ford B, Chiriboga CA, Clayton PT, Hinton V, Hyland K, Sharma R, De Vivo DC: **Aromatic L-amino acid decarboxylase deficiency: clinical features, treatment, and prognosis.** *Neurology* 2004, **62**(7):1058-1065.

220. Doehring A, Antoniades C, Channon KM, Tegeder I, Lotsch J: **Clinical genetics of functionally mild non-coding GTP cyclohydrolase 1 (GCH1) polymorphisms modulating pain and cardiovascular risk.** *Mutat Res* 2008, **659**(3):195-201.

221. Zelcer N, Hong C, Boyadjian R, Tontonoz P: **LXR regulates cholesterol uptake through Idol-dependent ubiquitination of the LDL receptor.** *Science* 2009, **325**(5936):100-104.

222. Vargas NB, Brewer BY, Rogers TB, Wilson GM: **Protein kinase C activation stabilizes LDL receptor mRNA via the JNK pathway in HepG2 cells.** *J Lipid Res* 2009, **50**(3):386-397.

223. Jeong JH, Cho S, Pak YK: **Sterol-independent repression of low density lipoprotein receptor promoter by peroxisome proliferator activated receptor gamma coactivator-1alpha (PGC-1alpha).** *Exp Mol Med* 2009, **41**(6):406-416.

224. Crooke RM, Graham MJ, Lemonidis KM, Whipple CP, Koo S, Perera RJ: **An apolipoprotein B antisense oligonucleotide lowers LDL cholesterol in hyperlipidemic mice without causing hepatic steatosis.** *J Lipid Res* 2005, **46**(5):872-884.

225. Hu J, Zhang Z, Shen WJ, Azhar S: **Cellular cholesterol delivery, intracellular processing and utilization for biosynthesis of steroid hormones.** *Nutr Metab (Lond)* 2010, **7**:47.

226. Perret B, Mabile L, Martinez L, Terce F, Barbaras R, Collet X: **Hepatic lipase: structure/function relationship, synthesis, and regulation.** *J Lipid Res* 2002, **43**(8):1163-1169.

227. Dichek HL, Qian K, Agrawal N: **The bridging function of hepatic lipase clears plasma cholesterol in LDL receptor-deficient "apoB-48-only" and "apoB-100-only" mice.** *J Lipid Res* 2004, **45**(3):551-560.

228. Maxwell KN, Soccio RE, Duncan EM, Sehayek E, Breslow JL: **Novel putative SREBP and LXR target genes identified by microarray analysis in liver of cholesterol-fed mice.** *J Lipid Res* 2003, **44**(11):2109-2119.

229. Beigneux AP, Kosinski C, Gavino B, Horton JD, Skarnes WC, Young SG: **ATP-citrate lyase deficiency in the mouse.** *J Biol Chem* 2004, **279**(10):9557-9564.

230. Ohgami M, Takahashi N, Yamasaki M, Fukui T: **Expression of acetoacetyl-CoA synthetase, a novel cytosolic ketone body-utilizing enzyme, in human brain.** *Biochem Pharmacol* 2003, **65**(6):989-994.

231. Mitchell ED, Jr., Avigan J: **Control of phosphorylation and decarboxylation of mevalonic acid and its metabolites in cultured human fibroblasts and in rat liver in vivo.** *J Biol Chem* 1981, **256**(12):6170-6173.

232. Breitling R, Laubner D, Clizbe D, Adamski J, Krisans SK: **Isopentenyl-diphosphate isomerases in human and mouse: evolutionary analysis of a mammalian gene duplication.** *J Mol Evol* 2003, **57**(3):282-291.

233. Fogarty MP, Xiao R, Prokunina-Olsson L, Scott LJ, Mohlke KL: **Allelic expression imbalance at high-density lipoprotein cholesterol locus MMAB-MVK.** *Hum Mol Genet* 2010, **19**(10):1921-1929.

234. Do R, Kiss RS, Gaudet D, Engert JC: **Squalene synthase: a critical enzyme in the cholesterol biosynthesis pathway.** *Clin Genet* 2009, **75**(1):19-29.

235. Crameri A, Biondi E, Kuehnle K, Lutjohann D, Thelen KM, Perga S, Dotti CG, Nitsch RM, Ledesma MD, Mohajeri MH: **The role of seladin-1/DHCR24 in cholesterol biosynthesis, APP processing and Abeta generation in vivo.** *EMBO J* 2006, **25**(2):432-443.
236. Waterham HR, Koster J, Mooyer P, Noort Gv G, Kelley RI, Wilcox WR, Wanders RJ, Hennekam RC, Oosterwijk JC: **Autosomal recessive HEM/Greenberg skeletal dysplasia is caused by 3 beta-hydroxysterol delta 14-reductase deficiency due to mutations in the lamin B receptor gene.** *Am J Hum Genet* 2003, **72**(4):1013-1017.
237. Rodriguez C, Raposo B, Martinez-Gonzalez J, Llorente-Cortes V, Vilahur G, Badimon L: **Modulation of ERG25 expression by LDL in vascular cells.** *Cardiovasc Res* 2003, **58**(1):178-185.
238. Caldas H, Herman GE: **NSDHL, an enzyme involved in cholesterol biosynthesis, traffics through the Golgi and accumulates on ER membranes and on the surface of lipid droplets.** *Hum Mol Genet* 2003, **12**(22):2981-2991.
239. Ohnesorg T, Keller B, Hrabe de Angelis M, Adamski J: **Transcriptional regulation of human and murine 17beta-hydroxysteroid dehydrogenase type-7 confers its participation in cholesterol biosynthesis.** *J Mol Endocrinol* 2006, **37**(1):185-197.
240. Bennett MK, Seo YK, Datta S, Shin DJ, Osborne TF: **Selective binding of sterol regulatory element-binding protein isoforms and co-regulatory proteins to promoters for lipid metabolic genes in liver.** *J Biol Chem* 2008, **283**(23):15628-15637.
241. Wang XH, Tian Y, Guo ZJ, Fan ZP, Qiu de K, Zeng MD: **Cholesterol metabolism and expression of its relevant genes in cultured steatotic hepatocytes.** *J Dig Dis* 2009, **10**(4):310-314.
242. Puglielli L, Tanzi RE, Kovacs DM: **Alzheimer's disease: the cholesterol connection.** *Nat Neurosci* 2003, **6**(4):345-351.
243. Hynynen R, Suchanek M, Spandl J, Back N, Thiele C, Olkkonen VM: **OSBP-related protein 2 is a sterol receptor on lipid droplets that regulates the metabolism of neutral lipids.** *J Lipid Res* 2009, **50**(7):1305-1315.
244. Perttila J, Merikanto K, Naukkarinen J, Surakka I, Martin NW, Tanhuanpaa K, Grimard V, Taskinen MR, Thiele C, Salomaa V *et al*: **OSBPL10, a novel candidate gene for high triglyceride trait in dyslipidemic Finnish subjects, regulates cellular lipid metabolism.** *J Mol Med* 2009, **87**(8):825-835.
245. Attie AD, Krauss RM, Gray-Keller MP, Brownlie A, Miyazaki M, Kastelein JJ, Lusis AJ, Stalenhoef AF, Stoehr JP, Hayden MR *et al*: **Relationship between stearoyl-CoA desaturase activity and plasma triglycerides in human and mouse hypertriglyceridemia.** *J Lipid Res* 2002, **43**(11):1899-1907.
246. Schwarz M, Wright AC, Davis DL, Nazer H, Bjorkhem I, Russell DW: **The bile acid synthetic gene 3beta-hydroxy-Delta(5)-C(27)-steroid oxidoreductase is mutated in progressive intrahepatic cholestasis.** *J Clin Invest* 2000, **106**(9):1175-1184.
247. Baumgart E, Vanhooren JC, Fransen M, Marynen P, Puype M, Vandekerckhove J, Leunissen JA, Fahimi HD, Mannaerts GP, van Veldhoven PP: **Molecular characterization of the human peroxisomal branched-chain acyl-CoA oxidase: cDNA cloning, chromosomal assignment, tissue distribution, and evidence for the absence of the protein in Zellweger syndrome.** *Proc Natl Acad Sci U S A* 1996, **93**(24):13748-13753.
248. Kubena LF, Edrington TS, Harvey RB, Buckley SA, Phillips TD, Rottinghaus GE, Casper HH: **Individual and combined effects of fumonisin B1 present in Fusarium moniliforme culture material and T-2 toxin or deoxynivalenol in broiler chicks.** *Poult Sci* 1997, **76**(9):1239-1247.
249. Zhou TX, Chen YJ, Yoo JS, Huang Y, Lee JH, Jang HD, Shin SO, Kim HJ, Cho JH, Kim IH: **Effects of chitooligosaccharide supplementation on performance, blood characteristics, relative organ weight, and meat quality in broiler chickens.** *Poult Sci* 2009, **88**(3):593-600.

250. Chowdhury SR, Smith TK: **Effects of feeding blends of grains naturally contaminated with Fusarium mycotoxins on performance and metabolism of laying hens.** *Poult Sci* 2004, **83**(11):1849-1856.
251. Gallaher CM, Munion J, Hesslink R, Jr., Wise J, Gallaher DD: **Cholesterol reduction by glucomannan and chitosan is mediated by changes in cholesterol absorption and bile acid and fat excretion in rats.** *J Nutr* 2000, **130**(11):2753-2759.
252. Calabrese V, Giuffrida Stella AM, Calvani M, Butterfield DA: **Acetylcarnitine and cellular stress response: roles in nutritional redox homeostasis and regulation of longevity genes.** *J Nutr Biochem* 2006, **17**(2):73-88.
253. van der Leij FR, Huijkman NC, Boomsma C, Kuipers JR, Bartelds B: **Genomics of the human carnitine acyltransferase genes.** *Mol Genet Metab* 2000, **71**(1-2):139-153.
254. Gross DN, van den Heuvel AP, Birnbaum MJ: **The role of FoxO in the regulation of metabolism.** *Oncogene* 2008, **27**(16):2320-2336.
255. Beale EG, Hammer RE, Antoine B, Forest C: **Disregulated glyceroneogenesis: PCK1 as a candidate diabetes and obesity gene.** *Trends Endocrinol Metab* 2004, **15**(3):129-135.
256. Puigserver P, Rhee J, Donovan J, Walkey CJ, Yoon JC, Oriente F, Kitamura Y, Altomonte J, Dong H, Accili D *et al*: **Insulin-regulated hepatic gluconeogenesis through FOXO1-PGC-1alpha interaction.** *Nature* 2003, **423**(6939):550-555.
257. Taniguchi CM, Emanuelli B, Kahn CR: **Critical nodes in signalling pathways: insights into insulin action.** *Nat Rev Mol Cell Biol* 2006, **7**(2):85-96.
258. Kobayashi T, Deak M, Morrice N, Cohen P: **Characterization of the structure and regulation of two novel isoforms of serum- and glucocorticoid-induced protein kinase.** *Biochem J* 1999, **344 Pt 1**:189-197.
259. Maresca M, Mahfoud R, Garmy N, Fantini J: **The mycotoxin deoxynivalenol affects nutrient absorption in human intestinal epithelial cells.** *J Nutr* 2002, **132**(9):2723-2731.
260. Kanai Y, Segawa H, Miyamoto K, Uchino H, Takeda E, Endou H: **Expression cloning and characterization of a transporter for large neutral amino acids activated by the heavy chain of 4F2 antigen (CD98).** *J Biol Chem* 1998, **273**(37):23629-23632.
261. Gaccioli F, Huang CC, Wang C, Bevilacqua E, Franchi-Gazzola R, Gazzola GC, Bussolati O, Snider MD, Hatzoglou M: **Amino acid starvation induces the SNAT2 neutral amino acid transporter by a mechanism that involves eukaryotic initiation factor 2alpha phosphorylation and cap-independent translation.** *J Biol Chem* 2006, **281**(26):17929-17940.
262. Robbana-Barnat S, Loridon-Rosa B, Cohen H, Lafarge-Frayssinet C, Neish GA, Frayssinet C: **Protein synthesis inhibition and cardiac lesions associated with deoxynivalenol ingestion in mice.** *Food Addit Contam* 1987, **4**(1):49-56.
263. Pestka J, Zhou HR: **Toll-like receptor priming sensitizes macrophages to proinflammatory cytokine gene induction by deoxynivalenol and other toxicants.** *Toxicol Sci* 2006, **92**(2):445-455.
264. Hayashi F, Means TK, Luster AD: **Toll-like receptors stimulate human neutrophil function.** *Blood* 2003, **102**(7):2660-2669.
265. Amuzie CJ, Shinozuka J, Pestka JJ: **Induction of suppressors of cytokine signaling by the trichothecene deoxynivalenol in the mouse.** *Toxicol Sci* 2009, **111**(2):277-287.
266. Rico-Bautista E, Flores-Morales A, Fernandez-Perez L: **Suppressor of cytokine signaling (SOCS) 2, a protein with multiple functions.** *Cytokine Growth Factor Rev* 2006, **17**(6):431-439.
267. Wong SS, Zhou HR, Pestka JJ: **Effects of vomitoxin (deoxynivalenol) on the binding of transcription factors AP-1, NF-kappaB, and NF-IL6 in raw 264.7 macrophage cells.** *J Toxicol Environ Health A* 2002, **65**(16):1161-1180.
268. Casteel M, Nielsen C, Kothlow S, Dietrich R, Martlbauer E: **Impact of DUSP1 on the apoptotic potential of deoxynivalenol in the epithelial cell line HepG2.** *Toxicol Lett* 2010.

269. Patterson KI, Brummer T, O'Brien PM, Daly RJ: **Dual-specificity phosphatases: critical regulators with diverse cellular targets**. *Biochem J* 2009, **418**(3):475-489.
270. Zhang S, Lin Y, Kim YS, Hande MP, Liu ZG, Shen HM: **c-Jun N-terminal kinase mediates hydrogen peroxide-induced cell death via sustained poly(ADP-ribose) polymerase-1 activation**. *Cell Death Differ* 2007, **14**(5):1001-1010.
271. Al-Ahmadi W, Al-Haj L, Al-Mohanna FA, Silverman RH, Khabar KS: **RNase L downmodulation of the RNA-binding protein, HuR, and cellular growth**. *Oncogene* 2009, **28**(15):1782-1791.
272. Wang Q, Nagarkar DR, Bowman ER, Schneider D, Gosangi B, Lei J, Zhao Y, McHenry CL, Burgens RV, Miller DJ *et al*: **Role of double-stranded RNA pattern recognition receptors in rhinovirus-induced airway epithelial cell responses**. *J Immunol* 2009, **183**(11):6989-6997.
273. Wache YJ, Hbabi-Haddioui L, Guzylack-Piriou L, Belkhelfa H, Roques C, Oswald IP: **The mycotoxin deoxynivalenol inhibits the cell surface expression of activation markers in human macrophages**. *Toxicology* 2009, **262**(3):239-244.
274. Krause CD, He W, Kotenko S, Pestka S: **Modulation of the activation of Stat1 by the interferon-gamma receptor complex**. *Cell Res* 2006, **16**(1):113-123.

7. Appendix A

Table 7.1: Selenium experiment; upregulated genes

Probe Set ID	Name	P-value	Fc	Description	Function
Transcription regulation					
Gga.7800.1.S1_at	LOC428660*	0.004	1.618	similar to very large inducible GTPase-1	Regulation of transcription
Gga.14966.1.S1_at	ARID4A*	0.040	1.499	AT rich interactive domain 4A (RBP1-like)	Negative regulation of transcription
GgaAffx.24729.1.S1_s_at	NCOA2*	0.047	1.307	nuclear receptor coactivator 2	Regulation of transcription
GgaAffx.24729.2.S1_s_at	NCOA2	0.031	1.302	nuclear receptor coactivator 2	Regulation of transcription
GgaAffx.20307.1.S1_s_at	HIF1A*	0.038	1.241	hypoxia-inducible factor 1, alpha subunit	Transcription factor
GgaAffx.1220.1.S1_at	PHTF1*	0.040	1.206	putative homeodomain transcription factor 1	Regulation of transcription
Signaling					
Gga.6962.2.S1_s_at	GPR137B	0.044	1.439	G protein-coupled receptor 137B	Multi-pass membrane protein
Gga.960.1.S1_at	VIPR1	0.002	1.290	vasoactive intestinal peptide receptor 1	G-protein coupl. recep. protein sig.
Gga.5847.1.S1_at	OXT	0.022	1.241	oxytocin, prepro- (neurophysin I)	G-protein coupl. recep. protein sig.
Gga.4070.1.S1_at	CSF1R	0.043	1.230	colony stimulating factor 1 receptor	Cytokine
GgaAffx.5086.1.S1_at	LOC422347*	0.047	1.437	similar to insulin receptor substrate protein	Possible insulin signaling
Gga.1065.1.S1_at	GNAI2*	0.018	1.227	guanine nucleotide binding protein (G protein), α	G-protein coupl. recep. protein sig.
Cell cycle / regulation					
Gga.11828.1.S1_at	NINJ1*	0.021	1.565	ninjurin 1	Cell adhesion
Gga.4520.1.S1_s_at	ACTG1	0.032	1.437	actin, gamma 1	Cell motility
Gga.18990.1.S1_at	BNIP1*	0.008	1.397	BCL2/adenovirus E1B 19kDa interacting protein 1	Apoptosis
GgaAffx.20228.1.S1_s_at	SHROOM3*	0.018	1.374	shroom family member 3	Cytoskeleton
Gga.4821.1.S1_at	COL18A1	0.030	1.362	collagen, type XVIII, alpha 1	Cell adhesion
GgaAffx.11741.1.S1_at	IHPK2	0.049	1.362	inositol hexaphosphate kinase 2	Negative regulation of cell growth
GgaAffx.20199.1.S1_s_at	RQCD1*	0.050	1.344	RCD1 required for cell differentiation1 homolog	Differentiation
Gga.4116.1.S1_at	CASP9	0.048	1.304	caspase 9, apoptosis-related cysteine peptidase	Apoptotic program
Gga.7471.2.A1_s_at	HSPG2*	0.042	1.302	heparan sulfate proteoglycan 2	Cell adhesion
GgaAffx.8013.1.S1_at	MTERFD3*	0.040	1.294	MTERF domain containing 3	Cell growth
Gga.18988.1.S1_at	MADD*	0.035	1.293	MAP-kinase activating death domain	Regulation of apoptosis / cell cycle
GgaAffx.22596.1.S1_s_at	EHMT1*	0.027	1.284	euchromatic histone methyltransferase 1	Chromatin modification
Gga.1294.1.S1_at	MYO1F*	0.040	1.238	myosin IF	Myosin complex
Gga.2597.1.S1_at	SMARCA1	0.024	1.218	SWI/SNF related, matrix associated, actin depend.	ATP-dependent chromatin remod.
Immune system					
Gga.1408.1.S1_at	C4BPA	0.023	1.302	complement component 4 binding protein, alpha	Complement activation
Gga.4414.3.S1_a_at	CD74	0.038	1.299	CD74 antigen	Immune response
GgaAffx.21845.2.S1_s_at	CD74	0.039	1.252	CD74 antigen	Immune response
Gga.17397.1.S1_at	MYD88	0.049	1.247	myeloid differentiation primary response gene (88)	Inflammatory response
GgaAffx.6854.1.S1_s_at	C8A	0.038	1.235	complement component 8, alpha polypeptide	Complement activation
Gga.1328.1.S2_at	B2M	0.021	1.206	beta-2-microglobulin	Immune response
Metabolism					
Gga.5691.1.S1_at	LOC417943*	0.042	2.041	similar to Osteocalcin precursor	Regulation of bone mineralization
GgaAffx.21755.1.S1_s_at	OGCHI	0.003	1.595	ovoglycoprotein	Glycoprotein
Gga.16883.1.S1_at	LONRF2*	0.016	1.506	LON peptidase N-terminal domain and ring finger 2	ATP-dependent proteolysis
Gga.8056.1.S1_at	SENP5*	0.048	1.456	SUMO1/sentrin specific peptidase 5	Ubiquitin cycle
Gga.5588.1.S1_at	PDZK1IP1*	0.042	1.414	PDZK1 interacting protein 1	unknown
Gga.15912.1.S1_at	CLUAP1*	0.040	1.412	clusterin associated protein 1	Protein binding
Gga.10420.2.S1_a_at	CSRP2BP*	0.041	1.412	CSRP2 binding protein	Metabolic process
Gga.2334.1.S2_at	PHOSPHO1	0.048	1.410	phosphatase, orphan 1	Metabolic process
GgaAffx.12986.1.S1_s_at	AARS2*	0.050	1.403	alanyl-tRNA synthetase 2, mitochondrial	alanyl-tRNA aminoacylation
Gga.4756.1.S1_at	CIRBP	0.021	1.325	cold inducible RNA binding protein	Response to cold
GgaAffx.21002.1.S1_s_at	ADAT1*	0.041	1.304	adenosine deaminase, tRNA-specific 1	tRNA processing
GgaAffx.7230.1.S1_s_at	BCS1L*	0.046	1.289	BCS1-like (yeast)	Protein complex assembly
GgaAffx.4487.1.S1_s_at	PGS1*	0.045	1.280	phosphatidylglycerophosphate synthase 1	Phospholipid biosynthetic process
Gga.999.2.S1_s_at	PPP2R4	0.049	1.256	protein phosphatase 2A activator, subunit 4	Protein AA dephosphorylation

Gga.15805.1.S1_at	VEGFR3*	0.025	1.255	fms-related tyrosine kinase 4	VEGF receptor activity
GgaAffx.2440.3.S1_s_at	PIGT*	0.024	1.253	phosphatidylinositol glycan anchor biosynthesis	Glycan-structures biosynthesis
GgaAffx.20992.1.S1_s_at	PSMB7*	0.040	1.241	proteasome subunit, beta type, 7	Protein degradation
Gga.4108.4.S1_x_at	TPM1	0.046	1.212	tropomyosin 1 (alpha)	Regulation of muscle contraction
Selenoproteins					
Gga.4896.1.S1_at	GPX1	0.009	2.924	similar to glutathione peroxidase (EC 1.11.1.9)	Response to oxidative stress
Gga.3109.1.A1_s_at	SEPP	0.001	2.045	selenoprotein P, plasma, 1	Response to oxidative stress
Gga.7044.1.S1_s_at	SEPP	0.001	1.972	selenoprotein P, plasma, 1	Response to oxidative stress
GgaAffx.9121.1.S1_at	APOA5	0.005	1.855	apolipoprotein A-V	Selenoprotein
Gga.107.1.S1_at	GPX4	0.032	1.429	glutathione peroxidase 4	Response to oxidative stress
Gga.7044.3.S1_s_at	SEPP	0.030	1.397	selenoprotein P, plasma, 1	Response to oxidative stress
GgaAffx.11541.1.S1_at	SEPP	0.038	1.372	selenoprotein P, plasma, 1	Response to oxidative stress
Gga.16436.1.S1_at	LOC424593	0.010	1.297	similar to selenoprotein Pb	Antioxidant defense
Transport					
GgaAffx.10869.1.S1_at	SLC25A30*	0.002	1.988	solute carrier family 25, member 30	Mitoch. transport antioxidant correl.
GgaAffx.24960.1.S1_at	PQLC3*	0.012	1.626	PQ loop repeat containing 3	Multi-pass membrane protein
GgaAffx.4637.1.S1_at	RHBDF1*	0.038	1.565	rhomboid 5 homolog 1 (Drosophila)	Transmembrane ER / Golgi
GgaAffx.12896.1.S1_at	SLC7A5*	0.021	1.520	solute carrier family 7, member 5	Amino acid transport
GgaAffx.26527.1.S1_at	TCOF1*	0.021	1.456	Treacher Collins-Franceschetti syndrome 1	Transport
GgaAffx.23508.1.S1_at	ATP2B1*	0.034	1.400	ATPase, Ca++ transporting, plasma membrane 1	Calcium transport
Gga.3329.1.S1_at	SLC20A2	0.039	1.399	solute carrier family 20, member 2	Phosphat transport
Gga.6149.1.S1_s_at	SLCO4A1*	0.048	1.395	SLC organic anion transporter family, member 4A1	Organic anion transporter
Gga.5221.2.S1_at	NUP98*	0.021	1.346	nucleoporin 98kDa	DNA replication
GgaAffx.2698.1.S1_s_at	ABCA8	0.039	1.318	ATP-binding cassette, sub-family A, member 8	Nucleotide binding / Transport
GgaAffx.23985.1.S1_at	NUP153*	0.040	1.299	nucleoporin 153kDa	mRNA transport
Gga.12475.1.S1_s_at	TTYH3*	0.028	1.267	tweety homolog 3 (Drosophila)	Ion transport
Unknown function					
Gga.1465.1.S1_at		0.036	2.141	Finished cDNA, clone ChEST75k8	
Gga.8288.1.S1_at	*	0.031	2.053	Transcribed locus	
Gga.9024.1.S1_at		0.050	1.894	Finished cDNA, clone ChEST478p5	
Gga.8176.1.S1_at		0.035	1.848	Transcribed locus	
GgaAffx.21928.1.S1_at	LOC415476*	0.048	1.764	similar to KIAA1199 protein	
Gga.15844.1.S1_s_at		0.025	1.712	Hypothetical protein, clone 14a17	
GgaAffx.13053.1.S1_s_at	LOC423625	0.050	1.701	similar to chromosome 10 open reading frame 58	
Gga.9397.1.S1_at	*	0.043	1.656	Transcribed locus	
Gga.6064.3.S1_a_at	LOC429268*	0.011	1.511	similar to FLJ00239 protein	
Gga.15951.1.S1_s_at	LOC421237*	0.036	1.499	similar to hypothalamus protein HT013	
Gga.1790.1.S1_at	*	0.044	1.488	Finished cDNA, clone ChEST322h21	
Gga.16820.1.S1_at	*	0.047	1.484	Finished cDNA, clone ChEST887l12	
Gga.16422.1.S1_at	*	0.038	1.468	Finished cDNA, clone ChEST377n6	
GgaAffx.23340.1.S1_at	LOC423423	0.008	1.456	hypothetical LOC423423	
Gga.7639.2.S1_a_at	*	0.013	1.453	LOC422001	
Gga.17677.1.S1_at		0.044	1.443	Finished cDNA, clone ChEST185n14	
Gga.2911.1.S1_at		0.022	1.443	Finished cDNA, clone ChEST926b22	
Gga.19280.1.S1_at	*	0.019	1.433	Finished cDNA, clone ChEST962e3	
Gga.10563.1.S1_at		0.040	1.427	Finished cDNA, clone ChEST198p6	
GgaAffx.20570.1.S1_s_at		0.027	1.425	Finished cDNA, clone ChEST1017a22	
Gga.4964.3.S1_x_at	LOC420760*	0.043	1.404	similar to T-cell receptor gamma Vg2-Jg2	
Gga.8775.1.S1_at		0.025	1.391	Finished cDNA, clone ChEST862e24	
GgaAffx.5489.1.A1_at		0.005	1.364	ENSGALT00000014188	
GgaAffx.8685.1.S1_s_at	LOC421005	0.013	1.362	similar to cDNA sequence BC005537	
Gga.8260.1.S1_at	LOC422000*	0.028	1.359	similar to putative protein product of HMFN0672	
Gga.16264.1.S1_at	LOC423978	0.048	1.357	hypothetical LOC423978	
Gga.7205.1.S1_at		0.036	1.357	Finished cDNA, clone ChEST731m12	
Gga.7207.1.S1_at	*	0.024	1.351	Transcribed locus	
Gga.17614.1.S1_at	*	0.033	1.349	Finished cDNA, clone ChEST378j19	
GgaAffx.21242.1.S1_at		0.028	1.328	Finished cDNA, clone ChEST883g17	
Gga.7676.1.S1_at		0.037	1.319	Transcribed locus	

Gga.8505.1.S1_at		0.041	1.318	Finished cDNA, clone ChEST921b19
GgaAffx.20985.1.S1_at	LOC421005	0.025	1.307	similar to cDNA sequence BC005537
Gga.9401.1.S1_at	LOC417343*	0.036	1.284	similar to hypothetical protein FLJ23825
GgaAffx.26486.1.S1_at		0.017	1.261	ENSGALT00000008629
Gga.16663.1.S1_s_at		0.044	1.241	Finished cDNA, clone ChEST371e1
Gga.16037.1.S1_at		0.022	1.232	Finished cDNA, clone ChEST64d13
Gga.4616.1.S1_at		0.036	1.218	Finished cDNA, clone ChEST218c17
Gga.5397.1.S1_s_at		0.036	1.212	Hypothetical protein, clone 33b23
GgaAffx.20868.1.S1_at		0.014	1.209	Finished cDNA, clone ChEST114a3

Fc: Fold change, GPRS: G-protein coupled receptor protein signaling pathway, AA: Amino acid, *: Raw value below 1000 (lowest raw value in at least one group 240 (without genes turned on).

Table 7.2: Selenium experiment; downregulated genes

Probe Set ID	Name	P-value	Fc	Description	Function
Transcription regulator					
GgaAffx.7254.1.S1_at	AZI2	0.016	1.757	5-azacytidine induced 2	Gene expression activator
Gga.2862.1.S1_at	NUDT6	0.020	1.690	nudix-type motif 6	Growth factor activity
Gga.4095.2.S1_a_at	PCAF	0.039	1.660	p300/CBP-associated factor	Basal transcription factor
Gga.5222.1.S1_at	ASXL1*	0.002	1.577	additional sex combs like 1 (Drosophila)	Regulation of transcription
GgaAffx.5634.1.S1_s_at	FOXN2*	0.010	1.537	forkhead box N2; synonym: HTLF	Regulation of transcription
GgaAffx.2844.1.S1_s_at	CUTL2*	0.044	1.527	cut-like 2 (Drosophila)	Regulation of transcription
GgaAffx.23312.1.S1_at	GLIS1	0.029	1.441	GLIS family zinc finger 1	Regulation of transcription
GgaAffx.10113.2.S1_s_at	CSDE1	0.007	1.321	cold shock domain containing E1, RNA-binding	Regulation of transcription
GgaAffx.13176.1.S1_s_at	NUCB2	0.009	1.311	nucleobindin 2	Transcription factor
Gga.8129.1.S1_at	ZNF511*	0.042	1.296	zinc finger protein 511	Regulation of transcription
Gga.6990.1.S1_s_at	SNAPC5*	0.034	1.287	small nuclear RNA activating compl. polypeptide 5	Regulation of transcription
Gga.16753.1.S1_s_at	TARDBP	0.044	1.274	TAR DNA binding protein	Regulation of transcription
Gga.4626.1.S1_at	GDI2	0.006	1.239	GDP dissociation inhibitor 2	Transcription regulation
GgaAffx.870.1.S1_s_at	LOC424740*	0.001	1.210	similar to Autoimmune regulator (APECED homol.)	Regulation of transcription
Transcription / Splicing					
Gga.8202.2.S1_a_at	ZFAND2B	0.025	1.543	zinc finger, AN1-type domain 2B	Metal ion binding
GgaAffx.10047.1.S1_at	MDN1*	0.036	1.479	MDN1, midasin homolog (yeast)	Reg. of protein complex assembly
Gga.13551.1.S1_at	EBNA1BP2	0.005	1.358	EBNA1 binding protein 2	Ribosome biogenesis and assembly
GgaAffx.26068.1.S1_at	HNRPH3	0.043	1.347	heterogeneous nuclear ribonucleoprotein H3 (2H9)	RNA splicing
GgaAffx.12855.1.S1_s_at	TSEN2*	0.015	1.345	tRNA splicing endonuclease 2 homolog	tRNA splicing
Gga.4833.1.S1_at	PAPD1	0.041	1.335	PAP associated domain containing 1	mRNA processing
GgaAffx.10509.2.S1_s_at	TAF2	0.023	1.332	TAF2 RNA polymerase II	Transcription initiation
GgaAffx.9134.1.S1_at	SEPSECS*	0.006	1.330	Sep tRNA:Sec tRNA synthase	Selenocysteine incorporation
Gga.8379.1.S1_at	FUSIP1	0.018	1.321	FUS interacting protein (serine/arginine-rich) 1	Regulation nuclear mRNA splicing
GgaAffx.13179.1.S1_s_at	TCEA1*	0.040	1.289	transcription elongation factor A (SII), 1	RNA elongation
GgaAffx.11586.1.S1_at	DDX19B	0.019	1.270	DEAD (Asp-Glu-Ala-As) box polypeptide 19B	mRNA export from nucleus
Gga.6313.1.S1_s_at	DUS2L	0.050	1.267	dihydrouridine synthase 2-like, SMM1 homolog	tRNA processing
Gga.19649.1.S1_s_at	SCLY*	0.015	1.256	selenocysteine lyase	Amino acid metabolic process
GgaAffx.22463.1.S1_at	MTIF2*	0.044	1.251	mitochondrial translational initiation factor 2	Regulation of translational initiation
Gga.6100.2.S1_a_at	LSM6	0.037	1.243	LSM6 homolog, U6 small nuclear RNA associated	RNA splicing
Gga.7559.1.S1_at	LOC415411	0.045	1.232	similar to CDNA sequence BC003885	Ribosome
Gga.12278.1.S1_a_at	DEADC1	0.033	1.231	deaminase domain containing 1	tRNA processing
Gga.4880.1.S1_s_at	HNRPH1	0.042	1.226	heterogeneous nuclear ribonucleoprotein H1 (H)	mRNA processing
Nucleic acid					
Gga.11684.1.S1_at	BXDC1*	0.048	1.490	brix domain containing 1	Nucleus protein
Gga.16934.1.S1_s_at	OBFC2A*	0.050	1.391	oligonucl./oligosacc.-binding fold containing 2A	Nucleic acid binding
Gga.3122.2.S1_a_at	REXO2	0.002	1.378	REX2, RNA exonuclease 2 homolog	Nucleotide metabolic process
GgaAffx.12159.1.S1_s_at	ZC3H15	0.007	1.349	zinc finger CCCH-type containing 15	Nucleic acid binding
Gga.8223.1.S1_s_at	TPMT	0.015	1.329	thiopurine S-methyltransferase	Nucleotide metabolic process
Translation					

Gga.4894.1.S1_s_at	NACA	0.031	1.517	nascent polypeptide-associated complex α subunit	Translation
Gga.1990.1.S1_at	MRPL35	0.018	1.339	mitochondrial ribosomal protein L35	Translation
Gga.7179.2.A1_a_at	MRPS18A	0.024	1.336	mitochondrial ribosomal protein S18A	Translation
Gga.16496.1.A1_at	NACA	0.002	1.289	nascent polypeptide-associated complex α subunit	Translation
Gga.15791.1.S1_s_at	EIF1AX	0.026	1.277	eukaryotic translation initiation factor 1A, X-linked	Translational initiation
Gga.15791.2.S1_s_at	EIF1AX	0.022	1.223	eukaryotic translation initiation factor 1A, X-linked	Translational initiation
Signal transduction					
GgaAffx.13136.1.S1_at	RAB33B	0.045	1.464	RAB33B, member RAS oncogene family	Small GTPase mediated sig.transd.
Gga.16576.1.S1_at	RIN2*	0.018	1.448	Ras and Rab interactor 2	Small GTPase mediated sig.transd.
Gga.7649.1.S1_at	GNB1L*	0.038	1.413	guanine nucleo. bind. protein, beta polypeptide 1-	G-prot. coupled recep. protein
Gga.6056.1.S1_at	MTDH	0.012	1.374	metadherin	Cell junction
GgaAffx.21143.1.S1_s_at	GPR89B*	0.047	1.371	G protein-coupled receptor 89B	G-prot. coupled recep. protein
Gga.1681.2.S1_a_at		0.047	1.318	moderately similar to transducin β-like 2 isoform 1	
Gga.13354.2.S1_a_at	GARNL3*	0.019	1.311	GTPase activating Rap/RanGAP domain-like 3	Reg. small GTPase mediat. sig.
Gga.464.1.S1_at	AXIN1*	0.024	1.294	axin 1	Wnt receptor signaling
Gga.4068.1.S2_at	RAN	0.023	1.288	RAN, member RAS oncogene family	Small GTPase mediated sig.transd.
Gga.4553.1.S1_at	RAB1A	0.041	1.280	RAB1A, member RAS oncogene family	Small GTPase mediated sig.transd.
GgaAffx.9920.1.S1_at	DEPDC2*	0.000	1.249	DEP domain containing 2	G-prot. coupled recep. protein
GgaAffx.12404.1.S1_s_at	GARNL3*	0.040	1.249	GTPase activating Rap/RanGAP domain-like 3	Reg. small GTPase mediat. sig.
Gga.11466.1.S1_at	GPR23*	0.045	1.210	G protein-coupled receptor 23	G-prot. coupled recep. protein
Gga.4948.1.S1_at	FKBP1A	0.011	1.204	FK506 binding protein 1A, 12kDa	Reg. I-kappaB kinase/NF-kB casc.
Protein modification					
Gga.3393.1.S1_at	GLDC	0.013	1.619	glycine dehydrogenase (decarboxylating)	Glycine cleavage
Gga.16381.1.S1_at	TTLL5	0.048	1.616	tubulin tyrosine ligase-like family, member 5	Protein modification
GgaAffx.7758.2.S1_s_at	METAP1*	0.003	1.382	methionyl aminopeptidase 1	N-terminal prot.AA modification
GgaAffx.1758.1.S1_s_at	DNAJC9	0.015	1.361	DnaJ (Hsp40) homolog, subfamily C, member 9	Chaperone
GgaAffx.11532.1.S1_s_at	CANX	0.044	1.337	calnexin	Chaperone
GgaAffx.11524.1.S1_s_at	CCT4	0.039	1.281	chaperonin containing TCP1, subunit 4 (delta)	Protein folding
Gga.2877.1.S1_at	DNAJC18*	0.041	1.281	DnaJ (Hsp40) homolog, subfamily C, member 18	Chaperone
GgaAffx.12333.1.S1_s_at	CCT5	0.045	1.272	chaperonin containing TCP1, subunit 5 (epsilon)	Protein folding
GgaAffx.11887.1.S1_at	ST13	0.018	1.263	suppression of tumorigenicity 13 (Hsp70 interact.)	Protein folding
GgaAffx.8270.1.S1_at	CCT4	0.012	1.260	chaperonin containing TCP1, subunit 4 (delta)	Protein folding
Gga.4812.1.S1_at	JAKMIP2	0.039	1.233	janus kinase and microtubule interacting protein 2	Inositol-1-monophosphatase activity
Ubiquitin proteolytic cycle					
Gga.10870.1.S1_a_at	ASB5	0.044	1.838	ankyrin repeat and SOCS box-containing 5	Ubiquitination
GgaAffx.12331.1.S1_at	RNF11	0.037	1.652	ring finger protein 11	Protein ubiquitination
Gga.13353.1.S1_at	PRSS23*	0.036	1.550	protease, serine, 23	Proteolysis
Gga.4218.1.S1_at	UFM1	0.014	1.529	ubiquitin-fold modifier 1	Ubiquitin cycle
Gga.16889.1.S1_at	MYLIP	0.040	1.507	myosin regulatory light chain interacting protein	Protein ubiquitination
GgaAffx.11480.1.S1_s_at	UFM1	0.029	1.492	ubiquitin-fold modifier 1	Ubiquitin cycle
Gga.10597.1.S1_at	USP24*	0.038	1.439	ubiquitin specific peptidase 24	Ubiquitin cycle
GgaAffx.13122.1.S1_s_at	POMP	0.006	1.430	proteasome maturation protein	Proteasome assebly
GgaAffx.270.1.S1_at	LOC426644*	0.039	1.408	similar to KIAA1815 protein	Proteolysis
Gga.8748.1.S1_s_at	FBXO22	0.012	1.392	F-box protein 22	Ubiquitin cycle
Gga.3670.2.S1_s_at	FBXO9	0.007	1.333	F-box protein 9	Protein ubiquitination
GgaAffx.24646.1.S1_at	SEC63	0.023	1.332	SEC63 homolog (S. cerevisiae)	Protein targeting to membrane
GgaAffx.11855.1.S1_s_at	PSMA2	0.042	1.330	proteasome subunit, alpha type, 2	Ubiquitin cycle
Gga.5999.3.S1_s_at	PSME3	0.040	1.325	proteasome activator subunit 3 (PA28 gamma; Ki)	Proteasome complex
GgaAffx.7188.2.S1_s_at	NGLY1	0.029	1.322	N-glycanase 1	Glycoprotein catabolic process
Gga.5765.1.S1_at	PSMA3	0.032	1.290	proteasome subunit, alpha type, 3	Ubiquitin cycle
GgaAffx.10043.1.S1_at	LOC427342*	0.049	1.287	similar to Topoisomerase I binding	Regulation of transcription
Gga.1071.1.S1_at	PSMC1	0.017	1.282	proteasome 26S subunit, ATPase, 1	Proteasome regulatory particle
GgaAffx.12061.1.S1_s_at	UBE2F	0.044	1.279	ubiquitin-conjugating enzyme E2F (putative)	Ubiquitin cycle
GgaAffx.5375.1.S1_s_at	XPNPEP1	0.045	1.271	X-prolyl aminopeptidase (aminopeptidase P) 1	Proteolysis
GgaAffx.8258.1.S1_at	MARCH6	0.037	1.261	membrane-associated ring finger (C3HC4) 6	Ubiquitin cycle
GgaAffx.22174.1.S1_at	HGS	0.040	1.246	hepatocyte growth fac.-regul.tyrosine kinase subst.	Signal transduction
GgaAffx.12849.1.S1_at	FBXO9	0.044	1.207	F-box protein 9	Ubiquitin cycle

GgaAffx.11938.1.S1_s_at	PSMD7	0.020	1.200	proteasome 26S subunit, non-ATPase, 7	Proteasome regulatory particle
Gga.11603.1.S1_at	CACYBP	0.019	1.200	calcyclin binding protein	Ubiquitin cycle
Response to stress / Immunesystem					
Gga.9580.2.S1_a_at	AKR1B1	0.031	3.526	aldo-keto reductase family 1, B1	Response to stress
GgaAffx.21796.1.S1_at	GAL13	0.002	2.300	beta-defensin 13	Defense response to bacterium
Gga.2533.1.S1_s_at	GSTA1	0.037	2.170	glutathione S-transferase A1	Glutathione metabolic process
GgaAffx.21433.1.S1_s_at	STK25*	0.036	1.605	serine/threonine kinase 25	Response to oxidative stress
GgaAffx.23371.1.S1_at	ANTXR2*	0.013	1.538	anthrax toxin receptor 2	Receptor activity
GgaAffx.12261.1.S1_s_at	NUDT1	0.033	1.490	nudix-type motif 1	Response to oxidative stress
Gga.2525.1.S1_s_at	GCLM	0.038	1.383	glutamate-cysteine ligase, modifier subunit	Glutathione synthesis
Gga.16007.1.S1_at	TDP1*	0.049	1.275	tyrosyl-DNA phosphodiesterase 1	DNA repair
Gga.5209.1.S2_at	PARP1	0.040	1.273	poly (ADP-ribose) polymerase family, 1	DNA repair
Gga.3883.2.S1_s_at	PRELID1	0.006	1.208	PRELI domain containing 1	Immune response
Metabolism					
Gga.12402.1.S1_at	R3HCC1	0.039	1.758	R3H domain and coiled-coil containing 1	Transferring alkyl or aryl groups
Gga.9212.1.S1_at	LOC422880*	0.005	1.672	nucleolar protein 14	Protein and RNA binding
Gga.12277.2.A1_at	ADH1C	0.047	1.605	alcohol dehydrogenase 1C (class I), γ	Alcohol dehydrogenase
Gga.9835.1.S1_a_at	TCTE1L	0.042	1.597	t-complex-associated-testis-expressed 1L	Bioloigical process
Gga.2184.1.S1_at	3.1	0.048	1.556	P311 POU	Biological process
Gga.17659.1.S1_at	BCHE*	0.017	1.553	butyrylcholinesterase	Beta-amyloid binding
Gga.2628.1.S1_s_at	ABAT	0.011	1.525	4-aminobutyrate aminotransferase	Alanine and aspartate metabolism
GgaAffx.11662.1.S1_s_at	GTPBP9	0.049	1.484	GTP-binding protein 9 (putative)	ATP catabolic process
Gga.8396.1.S1_s_at	DERA	0.036	1.457	2-deoxyribose-5-phosphate aldolase homolog	Pentose-phosphate pathway
Gga.8822.1.S1_at	NDUFB8	0.038	1.447	NADH dehydrogenase (ubiquinone) 1 beta	Electron transport
GgaAffx.22887.1.S1_s_at	GTPBP9	0.019	1.428	GTP-binding protein 9 (putative)	ATP catabolic process
Gga.11419.1.S1_at	HDHD1A*	0.045	1.425	haloacid dehalogenase-like hydrolase dom. cont. 1A	Metabolic process
Gga.1056.2.S1_s_at	GTPBP9*	0.022	1.424	GTP-binding protein 9 (putative)	ATP catabolic process
GgaAffx.24459.1.S1_s_at	MBOAT	0.047	1.420	memb. O-acyltransferase domain containing 5	Acyltransferase activity
Gga.10930.1.S1_at	LRRC54*	0.044	1.410	Tsukushin	Protein binding
Gga.5219.2.S1_a_at	LOC424109*	0.046	1.379	similar to alpha-aspartyl dipeptidase	Peptidase
Gga.11579.1.S1_at	LOC415661	0.017	1.368	similar to CG10964-PA	Metabolic process
Gga.5250.1.S1_a_at	HDHD3	0.035	1.352	haloacid dehalogenase-like hydrolase dom.cont. 3	Hydrolase activity
Gga.12675.1.S1_at	LOC427367*	0.021	1.349	similar to coiled-coil domain containing 2	Biological process
Gga.7342.1.S1_at	LOC427577*	0.022	1.340	similar to Glycerate kinase	Protein amino acid phosphorylation
GgaAffx.10650.2.S1_s_at	EXTL3	0.018	1.339	exostoses (multiple)-like 3	Heparan sulfate biosynthesis
Gga.9690.1.S1_at	LACTB2	0.009	1.326	lactamase, beta 2	Hydrolase activity
Gga.12574.1.S1_at	LOC418549	0.018	1.287	similar to pyridoxal kinase	Vitamine B6
GgaAffx.11848.1.S1_s_at	HIBADH	0.010	1.279	3-hydroxyisobutyrate dehydrogenase	Valine metabolic process
Gga.9722.1.S1_at	GALE	0.040	1.270	UDP-galactose-4-epimerase	Galactose metabolic process
Gga.3294.1.S1_at	SPCS3	0.038	1.263	signal peptidase complex subunit 3 homolog	Signal peptide processing
Gga.8284.1.S1_a_at	NDUFS6	0.046	1.252	NADH dehydrogenase (ubiquinone) Fe-S protein 6	Electron transport
Gga.12267.1.S1_s_at	NDUFAF1	0.028	1.252	NADH dehydrogenase (ubiquinone) 1 alpha	Electron transport
Gga.1829.1.S1_at	CCDC52*	0.011	1.250	coiled-coil domain containing 52	Biological process
Gga.8768.1.S1_at	PM5	0.032	1.221	pM5 protein	Carbohydrate binding
GgaAffx.9495.1.S1_at	GCNT4*	0.048	1.213	glucosaminyl (N-acetyl) transferase 4, core 2	Glycosylation
Lipid metabolism					
Gga.7847.1.S1_at	ACSL5	0.016	1.745	acyl-CoA synthetase long-chain family member 5	Fatty acid metabolism
Gga.5454.1.S1_at	ABHD5	0.032	1.729	abhydrolase domain containing 5	Lipid metabolic process
GgaAffx.12084.1.S1_s_at	ABHD5	0.029	1.596	abhydrolase domain containing 5	Lipid metabolic process
Gga.10239.1.S1_s_at	MTTP	0.048	1.545	microsomal triglyceride transfer protein	Lipid metabolic process
GgaAffx.12920.1.S1_at	ACSL5	0.040	1.520	acyl-CoA synthetase long-chain family member 5	Fatty acid metabolism
Gga.5952.1.S1_s_at	GPR175	0.004	1.372	G protein-coupled receptor 175	Lipid metabolic process
Gga.1242.3.S1_s_at	PITPNB	0.032	1.360	phosphatidylinositol transfer protein, beta	Lipid metabolic process
Gga.7347.2.S1_at	MCAT*	0.033	1.356	malonyl CoA:ACP acyltransferase (mitochondrial)	Fatty acid biosynthetic process
Gga.13371.1.S1_at	FADS1	0.050	1.332	fatty acid desaturase 1	Unsaturated fatty acid biosynthesis
GgaAffx.22677.1.S1_at	PNPLA7*	0.014	1.331	patatin-like phospholipase domain containing 7	Lipid metabolic process
GgaAffx.2598.1.S1_at	SOAT1*	0.012	1.289	sterol O-acyltransferase 1	Cholesterol metabolic process
GgaAffx.24935.1.S1_at	ELOVL5	0.024	1.277	elongation of long chain FA family member 5	Fatty acid biosynthetic process
Apoptosis					
Gga.10151.1.S1_at	PDCD6	0.009	1.633	programmed cell death 6	Apoptosis

GgaAffx.22940.1.S1_at	BRWD2*	0.014	1.374	bromodomain and WD repeat domain containing 2	Apoptosis / Proteolysis
GgaAffx.22512.1.S1_s_at	PDCD11*	0.040	1.351	programmed cell death 11	tRNA processing
GgaAffx.13220.1.S1_s_at	AIFM1	0.046	1.334	apoptosis-inducing factor, mitochondrion-asso. 1	Apoptosis
Gga.9020.1.S1_at	ATN1*	0.023	1.319	atrophin 1	Cell death
Gga.6987.1.S1_at	PDCD5	0.041	1.217	programmed cell death 5	Apoptosis
Structural proteins					
GgaAffx.22337.3.S1_s_at	MYO1B	0.044	1.616	myosin IB	Actin binding
Gga.2759.1.S1_at	ADD3*	0.032	1.503	adducin 3 (gamma)	Assembly spectrin / actrin
GgaAffx.8313.1.S1_s_at	EPS8*	0.039	1.470	epidermal growth factor recept. path. substrate 8	Actin remodeling
Gga.12424.1.S1_at	KIF21A	0.002	1.444	kinesin family member 21A	Microtubule-based movement
Gga.7794.1.S1_at	TUBB6*	0.033	1.444	tubulin, beta 6	Microtubule-based movement
Gga.11453.1.S1_at	TMOD3*	0.050	1.343	tropomodulin 3 (ubiquitous)	Actin binding
GgaAffx.5725.1.S1_s_at	ACTL6A*	0.016	1.332	actin-like 6A	Microtubule-based movement
Gga.13363.1.S1_at	MEMO1	0.006	1.310	mediator of cell motility 1	Cell motility
Gga.4737.1.S1_at	LOC415296	0.020	1.272	actin, gamma 1	Cell motility
Gga.19459.1.S1_s_at	CHST10*	0.047	1.269	carbohydrate sulfotransferase 10	Cell adhesion
Gga.7589.1.S1_at	ACTB*	0.028	1.231	Transcribed locus, strongly similar beta actin	Cell motility
Transport					
Gga.13311.1.S1_s_at	VPS26B*	0.010	4.005	vacuolar protein sorting 26 homolog B (S. pombe)	Protein transport
Gga.15124.1.S1_at	MOBKL1A*	0.004	1.829	MOB1, Mps One Binder kinase activator-like 1A	Prot. amino acid
GgaAffx.7202.1.S1_at	CRELD2	0.009	1.725	cysteine-rich with EGF-like domains 2	Transprort regulator
Gga.9023.2.S1_a_at	SNX4	0.007	1.710	sorting nexin 4	Endocytosis
GgaAffx.4710.1.S1_at	LOC423141*	0.027	1.680	similar to two pore segment channel 2	Ion transport
GgaAffx.26601.1.S1_s_at	ARFGAP1*	0.046	1.671	ADP-ribosylation factor GTPase activ. protein 1	Vesicle-mediated transport
GgaAffx.6963.1.S1_at	KCNK1*	0.014	1.545	potassium channel, subfamily K, member 1	Potassium ion transport
GgaAffx.1153.1.S1_s_at	CBWD1	0.037	1.541	COBW domain containing 1	Protein binding
Gga.7879.1.S1_a_at	PEX10	0.016	1.508	peroxisome biogenesis factor 10	Peroxisome organization/biogene
Gga.16002.1.S1_at	ENOX1*	0.026	1.459	ecto-NOX disulfide-thiol exchanger 1	Transprort protein
Gga.6038.1.S1_at	LOC417856*	0.047	1.430	similar to TMEM19 protein	Transmembrane protein
Gga.1035.2.S1_a_at	LMAN1	0.037	1.425	lectin, mannose-binding, 1	ER to Golgi vesicle-med. transport
GgaAffx.11900.1.S1_s_at	VPS53*	0.025	1.408	vacuolar protein sorting 53 homolog (S. cerevisiae)	Protein transport
Gga.3957.2.S1_a_at	ENAH	0.038	1.379	enabled homolog (Drosophila)	Intracellular transport
GgaAffx.23610.3.S1_at	SLC22A1	0.040	1.377	solute carrier family 22 (organic cation transporter)	Transport protein
GgaAffx.11445.1.S1_at	STX7	0.020	1.358	syntaxin 7	Post-Golgi vesicle-mediated
GgaAffx.4455.1.S1_at	KCTD6*	0.029	1.350	potassium channel tetramerisation dom.containing 6	Potassium ion transport
GgaAffx.12938.3.S1_at	MFN1	0.050	1.349	mitofusin 1	Mitochondrial fusion
GgaAffx.11520.1.S1_s_at	AP1M1	0.038	1.336	adaptor-related protein complex 1, mu 1 subunit	Vesicle-mediated transport
GgaAffx.21454.1.S1_s_at	CPNE3*	0.045	1.321	copine III	Vesicle-mediated transport
GgaAffx.24058.1.S1_at	IPO8*	0.023	1.311	importin 8	Intracellular protein transport
GgaAffx.12838.1.S1_s_at	ERP29	0.017	1.307	endoplasmic reticulum protein 29	Intracellular protein transport
GgaAffx.575.1.S1_at	LOC415883	0.046	1.306	similar to LOC414611 protein	Integral to membrane
GgaAffx.10904.1.S1_at	VPS36*	0.045	1.300	vacuolar protein sorting 36 homolog (S. cerevisiae)	Protein transport
Gga.12868.1.S1_at	BCCIP	0.046	1.294	BRCA2 and CDKN1A interacting protein	DNA repair
Gga.18955.1.S1_at	VPS24	0.014	1.287	vacuolar protein sorting 24 homolog (S. cerevisiae)	Protein transport
Gga.9958.1.S1_at	BET1	0.033	1.279	BET1 homolog (S. cerevisiae)	ER to Golgi vesicle-med. transport
Gga.3599.1.S1_at	SEC23A	0.048	1.278	Sec23 homolog A (S. cerevisiae)	Vesicle-mediated transport
Gga.8101.3.S1_at	TOMM7	0.042	1.250	translocase of outer mitochondrial membrane 7	Intracellular protein transport
GgaAffx.7232.2.S1_s_at	SLC44A5*	0.006	1.243	solute carrier family 44, member 5	Choline transporter like
Gga.4633.1.S1_at	VDAC3	0.049	1.240	voltage-dependent anion channel 3	Adenine transport
Gga.17007.1.S1_at	HIAT1	0.050	1.237	hippocampus abundant transcript 1	Transport
GgaAffx.11401.1.S1_at	STX2	0.041	1.223	syntaxin 2	Vesicle-mediated transport
Gga.16183.2.S1_s_at	SLC35A4*	0.044	1.218	solute carrier family 35, member A4	Carbohydrate transport
GgaAffx.1663.1.S1_at	GMPPB*	0.045	1.208	GDP-mannose pyrophosphorylase B	Fructose / mannose metabolism
Cell cycle / Growth					
GgaAffx.24692.1.S1_at	PDZK1*	0.013	1.966	PDZ domain containing 1	Cell proliferation
GgaAffx.24692.2.S1_s_at	PDZK1*	0.006	1.855	PDZ domain containing 1	Cell proliferation
Gga.10646.1.S1_at	LOC420724*	0.004	1.853	similar to CLIP-associating protein 1	Cell cycle
Gga.8194.1.S1_a_at	CENPQ	0.048	1.634	centromere protein Q	Chromosome
Gga.3200.1.S1_s_at	SMC4*	0.013	1.571	structural maintenance of chromosomes 4	Cell cycle
Gga.3823.2.S1_at	PER2*	0.048	1.564	period homolog 2 (Drosophila)	Circadian rhythm
Gga.421.2.S1_a_at	TK1*	0.039	1.519	thymidine kinase 1, soluble	DNA replication

Gga.16445.1.S1_a_at	CCDC5*	0.050	1.419	coiled-coil domain containing 5 (spindle associated)	Cell cycle
GgaAffx.3314.1.S1_s_at	TKT	0.035	1.409	transketolase (Wernicke-Korsakoff syndrome)	Regulation of growth
Gga.3302.1.S1_a_at	CEP192*	0.016	1.394	centrosomal protein 192kDa	Chromosome
Gga.2476.1.S1_at	HAT1	0.024	1.363	histone acetyltransferase 1	DNA packaging
Gga.1029.2.S1_a_at	RBBP4	0.015	1.356	retinoblastoma binding protein 4	Cell cycle
GgaAffx.1012.1.S1_at	ZZEF1*	0.036	1.341	zinc finger, ZZ-type with EF-hand domain 1	Regulation meta-/anaphase
GgaAffx.4141.1.S1_at	SPECC1L	0.036	1.338	SPECC1-like; Gallus gallus SPECC1-like	Cell cycle
GgaAffx.11551.1.S1_s_at	H2AFV	0.047	1.335	H2A histone family, member V	Nucleosome assembly
Gga.2986.2.S1_a_at	RB1*	0.042	1.315	retinoblastoma 1	Cell cycle
Gga.883.1.S1_at	RFC2	0.035	1.305	replication factor C (activator 1) 2, 40kDa	DNA replication
Gga.12862.1.S1_s_at	GNPTAB	0.020	1.295	N-acetylglucosamine-1-phosphate transferase α & β	Cell differentiation
Gga.3627.1.S1_at	PTPN3	0.028	1.283	protein tyrosine phosphatase, non-receptor type 3	Protein AA dephosphorylation
GgaAffx.8095.1.S1_at	GNPTAB*	0.046	1.239	N-acetylglucosamine-1-phosphate transferase α & β	Cell differentiation
Unknown function					
Gga.9618.1.S1_at		0.031	3.000	Finished cDNA, clone ChEST540c19	
Gga.5787.1.S1_at		0.025	2.887	Transcribed locus	
Gga.16590.1.S1_at	*	0.044	2.087	Finished cDNA, clone ChEST226a22	
Gga.12212.1.S1_a_at	*	0.019	1.957	Finished cDNA, clone ChEST26f22	
GgaAffx.21203.1.S1_at	*	0.036	1.873	Finished cDNA, clone ChEST692i4	
Gga.16917.1.S1_at		0.022	1.873	Finished cDNA, clone ChEST993l9	
GgaAffx.12402.1.S1_s_at		0.015	1.836	Hypothetical protein, clone 14g12	
GgaAffx.24252.1.S1_at	LOC430674*	0.002	1.806	similar to erythroid differentiation-related factor 1	
Gga.1334.1.S1_at		0.047	1.761	Transcribed locus	
Gga.16838.1.S1_at	*	0.024	1.750	Finished cDNA, clone ChEST490c22	
Gga.17132.1.S1_at		0.048	1.747	Finished cDNA, clone ChEST120n19	
Gga.12387.1.S1_at	LOC420238	0.021	1.745	plasma glutamate carboxypeptidase	
Gga.14371.1.S1_at		0.035	1.708	Finished cDNA, clone ChEST78p3	
Gga.7815.2.S1_a_at	LOC418170	0.034	1.693	similar to aldose reductase	
Gga.18470.1.S1_at	*	0.038	1.686	Finished cDNA, clone ChEST638m8	
Gga.8064.1.S1_at	*	0.045	1.642	Finished cDNA, clone ChEST151g8	
Gga.14030.1.S1_at		0.038	1.635	Finished cDNA, clone ChEST251j21	
Gga.1799.1.S1_x_at		0.018	1.632	Finished cDNA, clone ChEST443c18	
GgaAffx.25360.1.S1_at	EPB41L4A*	0.027	1.626	erythrocyte membrane protein band 4.1 like 4A	
Gga.2519.1.S1_at		0.020	1.625	Finished cDNA, clone ChEST741k1	
GgaAffx.20532.1.S1_s_at		0.030	1.611	Finished cDNA, clone ChEST587g22	
Gga.1322.1.S1_at		0.045	1.610	Finished cDNA, clone ChEST716c17	
GgaAffx.20200.1.S1_at		0.015	1.593	Finished cDNA, clone ChEST716k3	
Gga.2505.1.S1_s_at	LOC424064	0.007	1.584	similar to hypothetical protein FLJ38973	
Gga.18806.1.S1_at	*	0.029	1.577	Finished cDNA, clone ChEST292n17	
GgaAffx.25171.1.S1_at	MTMR6*	0.038	1.564	myotubularin related protein 6	
Gga.17155.1.S1_at		0.019	1.562	Finished cDNA, clone ChEST559i23	
Gga.2907.2.S1_at	LOC420983	0.016	1.560	similar to hypothetical protein FLJ20457	
GgaAffx.20532.1.S1_at	*	0.044	1.555	Finished cDNA, clone ChEST587g22	
GgaAffx.6669.1.S1_at	LOC421465	0.044	1.550	hypothetical LOC421465	
Gga.19347.1.S1_at	*	0.049	1.546	Finished cDNA, clone ChEST651i1	
GgaAffx.21043.1.S1_at	*	0.041	1.542	Finished cDNA, clone ChEST859d15	
Gga.2152.1.S2_at	LOC395762	0.048	1.537	KS5 protein	
Gga.13201.1.S1_at		0.013	1.532	Finished cDNA, clone ChEST376o23	
GgaAffx.1527.1.S1_at	LOC427076*	0.045	1.524	similar to KIAA1913 protein	
Gga.17403.1.S1_at		0.015	1.501	Finished cDNA, clone ChEST724o21	
Gga.6175.1.S1_s_at		0.016	1.499	Hypothetical protein, clone 37b22	
Gga.17478.1.S1_at		0.042	1.498	Finished cDNA, clone ChEST792k17	
GgaAffx.5477.1.S1_at	TRABD	0.017	1.493	TraB domain containing	
Gga.7490.1.S1_at		0.023	1.487	Finished cDNA, clone ChEST638l5	
Gga.7090.1.S1_s_at		0.045	1.482	Finished cDNA, clone ChEST49g17	
Gga.7688.3.S1_at		0.026	1.474	Hypothetical protein, clone 7e2	
GgaAffx.21654.1.S1_at		0.010	1.471	Finished cDNA, clone ChEST739f12	
GgaAffx.20851.1.S1_at		0.044	1.468	Finished cDNA, clone ChEST724j17	
Gga.1945.1.S1_at	LOC419015	0.013	1.456	hypothetical LOC419015	
Gga.2062.1.S1_at		0.015	1.455	Transcribed locus	
GgaAffx.21526.1.S1_at		0.045	1.453	Finished cDNA, clone ChEST568l11	
Gga.19685.1.S1_at		0.035	1.450	Finished cDNA, clone ChEST836f20	
Gga.8226.1.S1_s_at	LOC431580	0.035	1.448	similar to Ran binding protein 11	

GgaAffx.20892.1.S1_at		0.038	1.443	Finished cDNA, clone ChEST959k23
Gga.9191.1.S1_at	*	0.015	1.436	Transcribed locus
Gga.6185.1.S1_a_at		0.037	1.435	Finished cDNA, clone ChEST588i20
Gga.16200.1.S1_at		0.008	1.434	Finished cDNA, clone ChEST69h24
GgaAffx.20697.1.S1_at		0.049	1.427	Finished cDNA, clone ChEST671a22
Gga.6361.1.S1_at		0.050	1.427	Transcribed locus
Gga.18551.1.S1_at	*	0.023	1.426	Finished cDNA, clone ChEST404e20
Gga.2035.1.S1_at	*	0.048	1.426	Transcribed locus
Gga.7056.1.S1_at		0.036	1.415	Finished cDNA, clone ChEST436k9
GgaAffx.8135.1.S1_at	LOC418110*	0.011	1.409	similar to KIAA1718 protein
GgaAffx.2708.1.S1_at	LOC419205*	0.040	1.407	similar to C20orf123
GgaAffx.24869.1.S1_at	LOC420306	0.039	1.406	similar to xPRL-3
Gga.6384.1.S1_at		0.006	1.402	Finished cDNA, clone ChEST445m7
Gga.12612.1.S1_at	LOC422650	0.038	1.398	hypothetical LOC422650
Gga.3209.3.S1_at		0.041	1.396	ENSGALG00000002710
Gga.16203.1.S1_at		0.031	1.396	Finished cDNA, clone ChEST773o7
Gga.11336.1.S1_at	LOC426740	0.021	1.391	similar to 4833424P18Rik protein
GgaAffx.12334.1.S1_at	LOC423142	0.033	1.390	similar to oral cancer overexpressed 1
Gga.13896.1.S1_at		0.040	1.388	Finished cDNA, clone ChEST595d11
Gga.7055.2.S1_a_at		0.017	1.385	Finished cDNA, clone ChEST307i11
Gga.8964.1.S1_at		0.049	1.381	Finished cDNA, clone ChEST1009h4
GgaAffx.11969.1.S1_at	TRIM8	0.045	1.374	tripartite motif-containing 8
Gga.11883.4.S1_s_at		0.031	1.373	Finished cDNA, clone ChEST655g24
Gga.1298.1.S1_at	LOC416463	0.013	1.371	similar to hypothetical protein MGC10911
Gga.4879.2.S1_at		0.034	1.368	ENSGALG00000011351
Gga.16131.1.S1_s_at	OTUD6B	0.013	1.365	OTU domain containing 6B
GgaAffx.21597.1.S1_s_at		0.043	1.365	Finished cDNA, clone ChEST794e4
GgaAffx.20772.1.S1_at		0.049	1.361	Finished cDNA, clone ChEST936b17
Gga.13822.1.S1_at	*	0.032	1.360	Finished cDNA, clone ChEST337a6
GgaAffx.23253.1.S1_s_at	LOC421465*	0.033	1.357	hypothetical LOC421465
Gga.14519.1.S1_at	*	0.032	1.356	Finished cDNA, clone ChEST70o22
GgaAffx.2036.1.S1_s_at	LOC419916	0.046	1.348	similar to Expressed sequence AI314976
Gga.7623.1.S1_at	LOC421433	0.045	1.346	similar to C6orf129 protein
GgaAffx.21165.1.S1_s_at	*	0.030	1.344	Finished cDNA, clone ChEST897d1
Gga.8048.1.S1_a_at		0.045	1.337	Finished cDNA, clone ChEST388a17
Gga.7003.1.S1_at		0.048	1.333	Finished cDNA, clone ChEST55d23
Gga.13288.1.S1_at		0.032	1.332	Finished cDNA, clone ChEST96k14
Gga.14870.1.S1_at		0.017	1.329	Finished cDNA, clone ChEST290o22
Gga.1641.1.A1_at		0.040	1.329	Transcribed locus
Gga.7716.2.S1_a_at	LOC417383	0.015	1.327	hypothetical gene supported by BX933931
Gga.6029.1.S1_at		0.039	1.327	Finished cDNA, clone ChEST903h7
Gga.16070.1.S1_s_at		0.044	1.323	Finished cDNA, clone ChEST250m1
Gga.9723.2.S1_a_at		0.042	1.322	Finished cDNA, clone ChEST618p17
GgaAffx.25211.1.S1_at	LOC419015	0.050	1.316	hypothetical LOC419015
Gga.2907.1.S1_s_at	LOC420983	0.006	1.311	similar to hypothetical protein FLJ20457
Gga.18113.1.S1_at		0.014	1.308	Finished cDNA, clone ChEST492l23
GgaAffx.2140.1.S1_at	LOC417412	0.002	1.299	similar to RIKEN cDNA 5730455P16 gene
Gga.12302.1.S1_at	IMPAD1	0.014	1.293	inositol monophosphatase domain containing 1
Gga.7035.1.S1_a_at		0.048	1.292	Finished cDNA, clone ChEST612e5
Gga.2574.1.S1_at		0.019	1.290	Finished cDNA, clone ChEST286k17
Gga.15300.1.S1_at		0.041	1.290	Finished cDNA, clone ChEST407a14
GgaAffx.2824.2.S1_s_at	LOC415416	0.006	1.288	similar to KIAA1370 protein
Gga.4600.1.S1_a_at		0.042	1.288	Finished cDNA, clone ChEST1014k22
GgaAffx.6483.2.S1_s_at		0.015	1.284	ENSGALG00000010281
GgaAffx.20999.1.S1_at		0.035	1.283	Finished cDNA, clone ChEST331n1
Gga.17190.1.S1_at		0.027	1.280	Finished cDNA, clone ChEST419g1
Gga.5315.1.S1_a_at		0.034	1.280	similar to RIKEN cDNA 1110008P14
Gga.15862.1.S1_at	LOC418241	0.024	1.278	similar to MGC53303 protein
Gga.17078.1.S1_at		0.022	1.273	Finished cDNA, clone ChEST320c14
GgaAffx.21454.1.S1_s_at	CURP	0.004	1.267	curly protein
GgaAffx.10482.1.S1_at	LOC420345	0.026	1.267	hypothetical LOC420345
Gga.17561.1.S1_at	LOC426097	0.044	1.267	hypothetical gene supported by CR353580;
Gga.16850.1.S1_at		0.049	1.264	Finished cDNA, clone ChEST169a21
Gga.18987.1.S1_at		0.045	1.264	Finished cDNA, clone ChEST561b17

GgaAffx.20379.1.S1_at	*	0.033	1.263	Finished cDNA, clone ChEST914o3
Gga.14472.1.S1_at	LOC417875	0.016	1.262	similar to FLJ21963 protein
Gga.9569.1.S1_at		0.001	1.262	Finished cDNA, clone ChEST772I16
Gga.9217.1.S1_at		0.050	1.261	ENSGALG00000013107
GgaAffx.7084.1.S1_s_at	TTC13	0.011	1.260	tetratricopeptide repeat domain 13
Gga.1926.1.S1_at		0.049	1.257	Finished cDNA, clone ChEST380b13
Gga.15628.1.S1_at		0.044	1.256	Finished cDNA, clone ChEST380I15
Gga.10683.1.S1_at		0.038	1.255	Finished cDNA, clone ChEST997f19
GgaAffx.12662.1.S1_s_at		0.044	1.254	Hypothetical protein, clone 20d15
GgaAffx.21235.1.S1_at		0.014	1.251	Finished cDNA, clone ChEST886k10
Gga.9711.1.S1_at		0.034	1.248	Similar to hypothetical protein FLJ38663
Gga.3507.1.S1_at		0.038	1.247	Finished cDNA, clone ChEST204m20
Gga.10418.1.S1_at		0.013	1.247	Finished cDNA, clone ChEST57j17
Gga.19153.1.S1_at		0.030	1.242	Finished cDNA, clone ChEST736p18
Gga.4538.2.S1_a_at	LOC417604	0.030	1.240	similar to RIKEN cDNA 0610009E20
Gga.5476.1.S1_at	LOC424830	0.005	1.240	similar to MO25 protein (CGI-66)
Gga.18841.1.S1_at		0.048	1.239	Finished cDNA, clone ChEST412m12
Gga.6982.1.S1_at	LOC415553	0.016	1.234	similar to Hypothetical protein FLJ11506
Gga.18967.1.S1_s_at	GABPA	0.045	1.233	GA binding protein transcription factor, α, 60kDa
Gga.11096.1.S1_s_at	LOC418439	0.015	1.233	similar to MSTP055
Gga.17197.1.S1_at		0.038	1.233	Finished cDNA, clone ChEST982i22
Gga.11346.1.S1_at		0.029	1.232	Finished cDNA, clone ChEST144m24
Gga.9004.1.S1_at		0.049	1.231	Finished cDNA, clone ChEST230g14
Gga.17192.1.S1_at		0.038	1.231	Finished cDNA, clone ChEST399n18
Gga.18836.1.S1_at		0.038	1.226	Finished cDNA, clone ChEST893o6
GgaAffx.11494.1.S1_s_at		0.007	1.223	Hypothetical protein, clone 1n18
Gga.8904.1.S1_at		0.009	1.222	Transcribed locus
Gga.7137.1.S1_at		0.008	1.214	Finished cDNA, clone ChEST503c13
Gga.9711.2.S1_s_at		0.026	1.214	similar to hypothetical protein FLJ38663
Gga.5918.1.A1_a_at		0.041	1.211	Finished cDNA, clone ChEST324m6
Gga.19211.1.S1_s_at		0.038	1.209	Finished cDNA, clone ChEST17k11
GgaAffx.21597.1.S1_at		0.020	1.209	Finished cDNA, clone ChEST794e4
Gga.7035.4.S1_a_at		0.004	1.204	Finished cDNA, clone ChEST612e5
GgaAffx.23270.3.S1_s_at	LOC429998	0.027	1.203	LOC429998

Fc: Fold change

*: Raw value below 1000 (lowest raw value in at least one group 240)

Table 7.3: Fatty acid analysis in the liver

Group	Average control g / 100g	Stdev.	Average Se-yeast g / 100g	Stdev.	T-test	Common name
C14:0	0.313	0.068	0.238	0.078	0.015	Myristic acid
C14:1	0.034	0.019	0.022	0.015	NS	Myristoleic acid
C15:0	0.047	0.005	0.055	0.007	0.001	Pentadecanoic acid
C16:0i[1]	0.163	0.025	0.212	0.051	0.003	
C16:0	20.105	1.921	16.878	2.732	0.002	Palmitic acid
C16:1aiso[1]	0.230	0.043	0.245	0.066	NS	
C16:1	1.095	0.570	0.793	0.366	NS	Palmitoleic acid
C17:0	0.181	0.036	0.209	0.032	0.049	Margaric acid
C18:0	22.433	1.855	22.420	2.052	NS	Stearic acid
C18:1n9	15.535	3.335	13.243	4.424	NS	Oleic acid
C18:1n7[1]	0.922	0.214	0.962	0.158	NS	
C18:1n5[1]	0.046	0.012	0.042	0.014	NS	
C18:2n6 c+t	15.200	1.367	17.125	1.918	0.007	Linoleic acid
C18:2n4[1,2]	0.038	0.005	0.043	0.006	0.024	
C18:3n6	0.088	0.018	0.095	0.021	NS	Gamma-linolenic acid
C18:3n3	5.137	0.466	5.954	1.744	NS	α-Linolenic acid
C18:4n3[2]	0.060	0.012	0.079	0.080	NS	Stearidonic acid
C18:4n6[2]	0.022	0.002	0.024	0.003	0.032	
C20:0	0.096	0.012	0.100	0.010	NS	Arachidic acid
C20:1n9	0.242	0.027	0.250	0.050	NS	Eicosenoic acid

C20:2n6	0.531	0.076	0.590	0.064	0.042	Eicosadienoic acid
C20:2x[1]	0.149	0.028	0.171	0.030	NS	
C20:3n6	1.051	0.133	1.131	0.195	NS	Dihomo-gamma-linolenic acid
C20:4n6	3.678	0.616	4.536	1.122	0.021	Arachidonic acid
C20:3n3	0.398	0.087	0.480	0.088	0.025	Eicosatrienoic acid
C20:4n3[2]	0.291	0.023	0.266	0.049	NS	Eicosatetraenoic acid
C20:5n3	5.210	0.769	5.756	1.494	NS	Eicosapentaenoic acid
C22:0	0.062	0.015	0.070	0.022	NS	Behenic acid
C22:1n9	0.022	0.004	0.025	0.005	0.047	Erucic acid
C23:0	0.206	0.037	0.238	0.035	0.032	Tricosylic acid
C22:4n6[3]	0.023	0.006	0.027	0.007	NS	Adrenic acid
C22:4n3[3]	0.045	0.009	0.054	0.009	0.020	
C22:5n6[3]	0.037	0.011	0.032	0.008	NS	Docosapentaenoic acid
C22:5n3[3]	2.431	0.508	2.762	0.596	NS	Tetracosapentaenoic acid
C24:0	3.756	0.822	4.727	1.486	0.046	Lignoceric acid
C22:6n3	0.037	0.009	0.040	0.017	NS	Docosahexaenoic acid
Squalene[1]	0.090	0.015	0.104	0.013	0.016	
Total	100		100			
SFA	47.361	1.758	45.148	2.740	0.020	
MFA	18.125	4.115	15.583	4.973	NS	
PUFA	34.424	3.200	39.165	4.852	0.006	
n-3	13.609	1.437	15.390	2.274	0.023	
n-6	20.629	2.086	23.561	2.971	0.007	
n-6/n-3	1.482	0.138	1.498	0.160	NS	
Δ5 desaturase n-6	3.481	0.307	4.001	0.711	0.026	
Δ5 desaturase n-3	17.795	1.997	21.663	6.051	0.029	
20:3n6/20:4n3	3.609	0.405	4.258	0.668	0.003	
20:4n6/20:5n3	0.706	0.119	0.788	0.122	NS	
ELOVL5	0.414	0.039	0.4	0.046	NS	
ACSL5: C18:0 / C16:0	1.108	0.187	1.328	0.288	0.017	
C16/18:2n6	1.323	0.240	0.986	0.282	0.004	
16:0/18:3n3	3.914	0.513	2.835	0.818	0.002	
C16/C18	0.896	0.148	0.753	0.158	0.024	
C20-22	18.353	2.526	21.360	4.621	0.046	

St. dev.: Standard deviation, SFA: Saturated fatty acid, MFA: Monounsaturated fatty acid, PUFA: Polyunsaturated fatty acid, NS: not significant, P > 0.05, FAME: Fatty acid methyl esters. Number of samples: Control 14, Se-yeast 12, ELOVL5: (C20:4n-3 + C22:5n-3 + C20:3n-6 + C22:4n-6)/ (C18:4n-3 + C20:5n-3 + C18:3n-6 + C20:4n-6).

[1]: Internal verification. [2]: According to Mondello [27] [3]: According to Hoffman [28]

8. Appendix B

Table 8.1: 1st Deoxynivalenol experiment; upregulated genes

Probe Set ID	Name	P-value	Fc	Description	Function
Upregulated in DON1, DON 2.5 and DON 5					
Gga.13478.3.S1_a_at		0.0001	7.092	Finished cDNA, clone ChEST485m4	Unknown
Gga.2919.1.S1_at	*	0.0030	2.268	Finished cDNA, clone ChEST492h15	Unknown
Gga.17617.1.S1_at		0.0009	2.045	Finished cDNA, clone ChEST617i22	Unknown
GgaAffx.10275.1.S1_at	SAMD12	0.0326	2.020	sterile alpha motif domain containing 12	Signaling
Gga.6844.1.S1_at	RPP38	0.0225	1.739	ribonuclease P/MRP 38kDa subunit	tRNA processing
Gga.9157.1.S1_at	COX19	0.0164	1.675	COX19 cytochrome c oxidase assembly homolog	Metal transport
Gga.19614.1.S1_at	PAPD4	0.0015	1.639	PAP associated domain containing 4	RNA polyadenylation
GgaAffx.20219.1.S1_at		0.0017	1.600	Finished cDNA, clone ChEST276b1	Unknown
Gga.18424.1.S1_at		0.0010	1.587	Finished cDNA, clone ChEST663n16	Unknown
Gga.8343.2.S1_at		0.0216	1.580	Transcribed locus	Unknown
Gga.7063.1.S1_at		0.0021	1.499	Finished cDNA, clone ChEST908o11	Unknown
Gga.2962.2.S1_s_at	LOC425455	0.0036	1.484	hypothetical gene supported by CR353498	Unknown
Gga.15505.1.S1_s_at	LOC425186	0.0037	1.481	similar to MGC80493 protein	Unknown
Gga.11720.1.S1_at	TTC36*	0.0309	1.418	tetratricopeptide repeat domain 36	Unknown
GgaAffx.11292.2.S1_s_at	LOC425649	0.0058	1.404	similar to rhomboid family 1	Unknown
Gga.3382.1.S1_at	PPIB*	0.0148	1.351	peptidylprolyl isomerase B (cyclophilin B)	Immune system
Gga.17085.1.S1_at	*	0.0179	1.319	Finished cDNA, clone ChEST544j6	Unknown
GgaAffx.12615.1.S1_s_at	*	0.0004	1.314	Hypothetical protein, clone 19f23	Unknown
GgaAffx.1528.1.S1_at	MAPKAPK3	0.0010	1.299	MAPK-activated protein kinase 3	p38 MAPK pathway
Gga.9906.1.S1_at	METRNL	0.0064	1.290	meteorin, glial cell differentiation regulator-like	Unknown
Gga.7226.1.S1_at	*	1.00E-04	1.282	Finished cDNA, clone ChEST140i2	Unknown
Gga.5306.2.S1_a_at		0.0237	1.256	Transcribed locus	Unknown
Upregulated in DON 2.5 and DON 5					
GgaAffx.24663.2.S1_s_at	IFT57	0.0044	1.953	intraflagellar transport 57 homolog	Regulation of Apoptosis
GgaAffx.4365.1.S1_at	LOC416968	0.0124	1.880	similar to presenilin-like protein 4	Peptidase
Gga.2529.2.S1_a_at	IL1RL1	0.0243	1.661	interleukin 1 receptor-like 1	Innate immune response
Gga.9700.2.S1_a_at	TSPAN7	0.0172	1.395	tetraspanin 7	Signal transduction
Upregulated in DON 1 and DON 5					
Gga.16554.2.S1_a_at	LAPTM4B*	0.0131	3.058	lysosomal associated protein transmembrane 4 β	Transmembrane protein
GgaAffx.2792.1.A1_at	LOC770379*	0.0022	2.646	hypothetical protein LOC770379	Unknown
Gga.8034.1.S1_at	LOC426712*	0.0052	2.421	hypothetical gene supported by CR387483	Unknown
Gga.10204.1.S1_s_at	CASP1	0.0056	2.198	caspase 1	Apoptosis
Gga.8730.1.S1_x_at	*	0.0202	2.037	Finished cDNA, clone ChEST414e17	Unknown
Gga.11819.1.S1_at	CENPQ	0.0315	1.828	centromere protein Q	Centromeric complex
Gga.19716.1.S1_at		0.0312	1.821	Finished cDNA, clone ChEST680j17	Unknown
Gga.11184.1.S1_at	KCTD12	0.0160	1.757	potassium channel tetramerisation domain cont. 12	Potassium ion transport
Gga.13063.1.S1_at		0.0047	1.757	Finished cDNA, clone ChEST534g17	Unknown
Gga.19225.1.S1_at		0.0115	1.742	Finished cDNA, clone ChEST668j19	Unknown
Gga.17784.1.S1_x_at		0.0204	1.730	Finished cDNA, clone ChEST887j4	Unknown
GgaAffx.10735.1.S1_at	MITD1	0.0052	1.672	MIT, microtubule interacting and transport, dom. 1	Protein transport
Gga.2952.1.S1_at	CLDN3	0.0310	1.637	claudin 3	Leukocyte transendothelial migration
Gga.13290.1.S1_at	RNFT1*	0.0023	1.582	ring finger protein, transmembrane 1	Metal ion binding
GgaAffx.12447.1.S1_at	SLC16A9	0.0021	1.570	solute carrier family 16, member 9	Monocarboxylic acid transporter 9)
GgaAffx.830.1.S1_at	LOC426958	0.0129	1.570	hypothetical LOC426958	Unknown
Gga.7669.1.S1_at	*	0.0211	1.548	Finished cDNA, clone ChEST435b7	Unknown
Gga.3364.2.S1_at	LOC425437*	0.0248	1.548	similar to phosphatidylinositol glycan, class B	Glycolipid mannosyltransferase activity
Gga.2252.1.S1_s_at	RAB33B*	0.0179	1.410	RAB33B, member RAS oncogene family	Unknown
GgaAffx.11111.1.S1_s_at	GDPD5	0.0167	1.410	glycerophosphodiester phosphodiesterase dom. 5	Glycerol metabolic process
GgaAffx.13080.1.S1_at	YPEL2*	0.0080	1.393	yippee-like 2 (Drosophila)	Cell division
GgaAffx.23739.1.S1_at	ARSJ	0.0007	1.350	arylsulfatase family, member J	Metabolic process
Gga.5367.1.S1_at	MALT1*	0.0008	1.316	mucosa assoc. lymphoid lymphoma transloc. 1	Defense response
GgaAffx.20734.1.S1_s_at	TMED10*	0.0104	1.297	transmembrane emp24-like trafficking protein 10	Vesicle-mediated transport
Gga.6785.1.A1_at	*	0.0118	1.279	similar to Dolichol-phosphate mannosyltransferase	Unknown
Gga.15821.1.S1_at	LASP1	0.0057	1.263	LIM and SH3 protein 1	Ion transport
Gga.13379.1.S1_at		0.0056	1.261	Finished cDNA, clone ChEST697k3	Unknown

GgaAffx.20300.1.S1_at	*	0.0276	1.232	Finished cDNA, clone ChEST816l12	Unknown
Upregulated in DON 1 and DON 2.5					
Gga.14721.1.S1_at		0.0029	2.475	Finished cDNA, clone ChEST125o17	Unknown
Gga.5708.2.S1_a_at	LOC415987	0.0026	2.309	similar to RIKEN cDNA 1110007C09	Unknown
Gga.8217.1.S1_at	*	0.0083	2.155	Finished cDNA, clone ChEST737p20	Unknown
GgaAffx.12452.1.S1_s_at	KLF11*	0.0150	2.092	Kruppel-like factor 11	Neg. reg. cell proliferation
Gga.14192.1.S1_at		0.0034	1.698	Finished cDNA, clone ChEST544b3	Unknown
Gga.13186.1.S1_at	*	0.0360	1.623	Finished cDNA, clone ChEST576c16	Unknown
Gga.19079.1.S1_at		0.0043	1.475	Finished cDNA, clone ChEST792o19	Unknown
Gga.16721.1.S1_at	LOC423369	0.0158	1.456	similar to RIKEN cDNA G630009D10 gene	Unknown
GgaAffx.11154.1.S1_at		0.0175	1.451	Transcribed locus	Unknown
Gga.18177.1.A1_at		0.0055	1.381	Finished cDNA, clone ChEST274p13	Unknown
GgaAffx.11322.1.S1_x_at	*	0.0329	1.344	Transcribed locus	Unknown
Gga.12469.1.S1_x_at	*	0.0117	1.300	Finished cDNA, clone ChEST647b6	Unknown
GgaAffx.11864.1.S1_at	GPR174	0.0089	1.300	G protein-coupled receptor 174	Signal transduction
GgaAffx.8019.1.S1_s_at	ATP9B	0.0160	1.272	ATPase, class II, type 9B	ATP biosynthetic process
Gga.3223.1.S1_at	YIPF6	0.0023	1.261	Yip1 domain family, member 6	Multi-pass membrane protein
Gga.6781.1.S1_at	*	0.0156	1.252	moderately similar to FK506 binding protein 11	Protein folding
GgaAffx.21176.1.S1_at		0.0307	1.203	Finished cDNA, clone ChEST763d5	Unknown
Gga.1723.1.S1_a_at	*	0.0364	1.200	Finished cDNA, clone ChEST153i21	Unknown
Upregulated in DON 5					
Gga.12857.1.S1_s_at	TP53I3	0.0245	2.646	tumor protein p53 inducible protein 3	Induction of apoptosis by oxidative
Gga.13986.1.S1_at	IFT57	0.0029	2.618	Intraflagellar transport 57 homolog	Regulation of Apoptosis
Gga.4124.1.S1_at	LOC395933*	0.0278	2.597	sulfotransferase	Sulfotransferase activity
Gga.6828.2.S1_a_at	LOC425419	0.0474	2.033	similar to rac GTPase activating protein	Unknown
Gga.4332.1.S1_at	HSPCB	0.0295	1.972	heat shock 90kDa protein 1, beta	Response to unfolded protein
Gga.635.1.S1_at	GLRX*	0.0312	1.901	glutaredoxin (thioltransferase)	Cell redox homeostasis
Gga.7555.1.S1_at	LOC425652	0.0197	1.894	similar to hypothetical protein FLJ20174	Unknown
GgaAffx.7092.1.S1_at	GNPAT*	0.0291	1.869	glyceronephosphate O-acyltransferase	Fatty acid metabolic process
GgaAffx.8046.1.S1_s_at	SLC41A2*	0.0316	1.832	solute carrier family 41, member 2	Cation transport
GgaAffx.12177.1.S1_s_at	MCM5	0.0222	1.832	minichromosome maintenance complex comp. 5	DNA replication, cell cycle
GgaAffx.24672.1.S1_s_at	PROS1*	0.0111	1.812	protein S (alpha)	Blood coagulation
Gga.3251.1.S1_at	NDUFS1	0.0128	1.802	NADH dehydrogenase Fe-S protein 1, 75kDa	ATP metabolic process
GgaAffx.12875.1.S1_at	VPS29	0.0366	1.779	vacuolar protein sorting 29 homolog (S. cerevisiae)	Protein transport
Gga.16579.1.S1_at		0.0321	1.776	Transcribed locus	Unknown
Gga.7164.1.S1_at	LOC421110	0.0252	1.724	similar to DNA replication initiator protein	DNA replication initiation
Gga.1190.1.S1_at	*	0.0019	1.712	Finished cDNA, clone ChEST633c17	Unknown
Gga.5235.1.S1_at	HYOU1*	0.0046	1.695	hypoxia up-regulated 1	Response to hypoxia
Gga.17818.1.A1_at		0.0167	1.684	Finished cDNA, clone ChEST886f3	Unknown
GgaAffx.5663.1.S1_at	SMAD7	0.0223	1.684	SMAD family member 7	Signaling
Gga.8338.1.S1_at	NIP7*	0.0472	1.667	nuclear import 7 homolog (S. cerevisiae)	Ribosome assembly
GgaAffx.21723.1.S1_at	GPSM1	0.0014	1.645	G-protein signaling modulator 1 (AGS3-like)	Cell differentiation
GgaAffx.4506.2.S1_s_at	RAB3IL1	0.0140	1.634	RAB3A interacting protein (rabin3)-like 1	Guanyl-nucleotide exchange factor
Gga.7843.1.S1_at	SYPL1*	0.0161	1.626	synaptophysin-like 1	Transport
Gga.18440.1.S1_at	*	0.0180	1.603	Finished cDNA, clone ChEST659p16	Unknown
Gga.14523.1.S1_at		0.0169	1.603	Finished cDNA, clone ChEST665e7	Unknown
Gga.11918.1.S1_at	NSUN3	0.0092	1.582	NOL1/NOP2/Sun domain family, member 3	Transferase activity
Gga.17784.1.S1_a_at		0.0283	1.577	Finished cDNA, clone ChEST887j4	Unknown
Gga.2119.2.A1_x_at	LOC426015	0.0068	1.572	hypothetical LOC426015	Unknown
GgaAffx.12146.1.S1_at	SNRPA1	0.0329	1.558	small nuclear ribonucleoprotein polypeptide A'	RNA splicing
GgaAffx.13032.1.S1_s_at	KIF23	0.0139	1.550	kinesin family member 23	Mitotic spindle elongation
Gga.16023.1.S1_s_at	LOC422442	0.0211	1.543	hypothetical gene supported by CR385915	Unknown
Gga.10691.1.S1_at	*	0.0399	1.538	Finished cDNA, clone ChEST271o11	Unknown
GgaAffx.12982.1.S1_s_at	COPB1*	0.0267	1.536	coatomer protein complex, subunit beta 1	Vesicle-mediated transport
Gga.6206.1.S1_s_at	CHMP6	0.0284	1.536	chromatin modifying protein 6	Protein transport
Gga.19911.1.S1_at		0.0385	1.531	Finished cDNA, clone ChEST801c15	Unknown
GgaAffx.8553.1.S1_s_at	PBX4	0.0415	1.524	pre-B-cell leukemia homeobox 4	Transcription regulation
GgaAffx.13133.1.S1_s_at	BPGM	0.0379	1.520	2,3-bisphosphoglycerate mutase	Glycolysis
GgaAffx.23087.1.S1_at	NUBPL	0.0367	1.515	nucleotide binding protein-like	Chromosome partitioning
Gga.8478.1.S1_s_at	KDELR2*	0.0288	1.513	KDEL ER protein retention receptor 2	Intracellular protein transport
Gga.8100.1.S1_a_at	*	0.0390	1.479	moderately similar to LOC68032	Unknown
Gga.3604.1.S1_at	UBR4*	0.0232	1.477	ubiquitin protein ligase E3 component n-recognin 4	Ubiquitin-protein ligase activity

Gga.14008.1.S1_at	ZNF236*	0.0434	1.473	Zinc finger protein 236	Regulation of transcription
Gga.7249.1.S1_a_at	CDT1	0.0236	1.473	chromatin licensing and DNA replication factor 1	Regulation of DNA replication initiation
GgaAffx.2959.1.S1_at	LOC420363	0.0120	1.473	similar to voltage-gated K^+ channel KV11.1	Transport
GgaAffx.20694.1.S1_s_at	ORMDL1*	0.0125	1.468	ORM1-like 1 (S. cerevisiae)	Protein folding
Gga.4473.1.S1_at		0.0296	1.464	Finished cDNA, clone ChEST995m24	Unknown
Gga.6864.1.S1_at	LONRF2	0.0208	1.462	LON peptidase N-terminal domain and ring finger 2	Proteolysis
GgaAffx.22254.1.S1_at	AMH	0.0020	1.460	anti-Mullerian hormone	Cell-Cell signaling
Gga.6610.1.S1_at	TMEM171	0.0235	1.458	Transmembrane protein 171	Membrane protein
GgaAffx.20999.1.S1_at		0.0228	1.447	Finished cDNA, clone ChEST331n1	Unknown
Gga.17168.1.S1_at	FBXL21*	0.0359	1.441	F-box and leucine-rich repeat protein 21	Protein ubiquitination
Gga.16676.1.S1_at		0.0163	1.439	Finished cDNA, clone ChEST212p21	Unknown
Gga.2136.3.S1_at		0.0169	1.437	strongly similar to ras homolog gene family, U	Signal transduction
Gga.4314.1.S1_at	P50	0.0178	1.435	dynamitin	Microtubule-based process
Gga.8862.1.S1_s_at	CHAF1B	0.0334	1.435	chromatin assembly factor 1, subunit B (p60)	DNA replication
GgaAffx.23168.1.S1_at	ALDH6A1*	0.0421	1.429	aldehyde dehydrogenase 6 family, member A1	Propanoate metabolism
Gga.1737.1.S1_at	SLC27A4*	0.0270	1.427	solute carrier family 27, member 4	Fatty acid transporter
Gga.6330.3.A1_at	RNF7	0.0299	1.425	ring finger protein 7	Ubiquitin cycle Apoptosis
Gga.2057.1.S1_at	LOC424657	0.0411	1.425	similar to hypothetical protein FLJ32112	Unknown
Gga.4774.1.S1_at	LOC422278*	0.0266	1.422	nonhistone chromosomal protein HMG-14A	DNA binding
Gga.19292.1.S1_at		0.0264	1.420	Finished cDNA, clone ChEST934j21	Unknown
Gga.229.1.S1_at	HINT1*	0.0246	1.414	histidine triad nucleotide binding protein 1	Signal transduction
Gga.5209.1.S2_at	PARP1*	0.0052	1.410	poly (ADP-ribose) polymerase family, member 1	Base excision repair
Gga.840.2.S1_a_at		0.0030	1.406	Myosin alkali light chain mRNA	Motor activity
GgaAffx.22246.1.S1_at	MPG	0.0313	1.406	N-methylpurine-DNA glycosylase	DNA repair
Gga.5731.1.S1_at	*	0.0158	1.403	Finished cDNA, clone ChEST752d2	Unknown
Gga.5951.1.S1_at		0.0231	1.403	Finished cDNA, clone ChEST973j19	Unknown
Gga.6994.1.S1_at	LOC431511*	0.0137	1.401	similar to Zgc:65827	Unknown
Gga.4694.1.S1_s_at	CCDC72*	0.0420	1.399	Coiled-coil domain containing 72	Apoptosis related
GgaAffx.12559.1.S1_s_at	SLC10A7	0.0301	1.399	solute carrier family 10, member 7	Ion transport
Gga.9929.1.S1_at	ELOVL1*	0.0380	1.395	elongation of very long chain fatty acids-like 1	Fatty acid biosynthetic process
GgaAffx.2464.1.S1_at	PRMT3	0.0365	1.395	protein arginine methyltransferase 3	Naphthalene/anthracene degradation
Gga.16283.1.S1_at		0.0321	1.391	Finished cDNA, clone ChEST354i6	Unknown
Gga.11633.1.S1_at		0.0092	1.389	Finished cDNA, clone ChEST1002p4	Unknown
Gga.12551.1.S1_at		0.0148	1.379	Finished cDNA, clone ChEST409o15	Unknown
Gga.11627.1.S1_at		0.0269	1.379	Transcribed locus	Unknown
GgaAffx.2506.2.S1_s_at	LOC416454*	0.0317	1.368	similar to cytochrome P450, family 2, W1	Xenobiotic metabolism
Gga.19163.1.S1_at	*	0.0198	1.366	Finished cDNA, clone ChEST993d13	Unknown
GgaAffx.26306.1.A1_s_at	*	0.0228	1.366	Transcribed locus	Unknown
GgaAffx.21403.1.S1_s_at	EME1	0.0479	1.366	essential meiotic endonuclease 1 homolog 1	DNA repair
Gga.5519.1.S1_at	TMEM183B*	0.0016	1.362	transmembrane protein 183B	Unknown
Gga.14483.1.A1_at		0.0457	1.361	Finished cDNA, clone ChEST327d20	Unknown
Gga.13259.1.S1_s_at	LIMD1*	0.0143	1.359	LIM domains containing 1	Repressor E2F-mediated transcrption
Gga.16516.1.S1_at		0.0348	1.353	Finished cDNA, clone ChEST136o22	Unknown
Gga.16053.1.S1_at	RCOR3	0.0259	1.351	REST corepressor 3	Regulation of transcription
Gga.10276.1.A1_at		0.0147	1.350	Finished cDNA, clone ChEST848l8	Unknown
Gga.9923.1.S1_s_at	LOC417353*	0.0343	1.344	hypothetical gene supported by BX933423	Unknown
Gga.4746.1.S1_a_at	NDUFB1*	0.0120	1.342	NADH dehydrogenase 1 beta subcomplex, 1, 7kDa	Oxidative phosphorylation
Gga.4839.1.S1_a_at	MDH2*	0.0325	1.342	malate dehydrogenase 2, NAD (mitochondrial)	Malate metabolic process
Gga.2811.1.S1_at		0.0055	1.342	Transcribed locus	Unknown
Gga.7381.2.S1_s_at	WDR45L	0.0421	1.340	WDR45-like	Related to autophagy
Gga.17504.1.S1_at	ILDR2	0.0018	1.340	immunoglobulin-like domain containing receptor 2	Cell differentiation
Gga.19435.1.A1_at		0.0026	1.339	Transcribed locus	Unknown
Gga.4600.1.S1_a_at	*	0.0295	1.337	Finished cDNA, clone ChEST1014k22	Unknown
Gga.6154.1.S1_at	TMEM59L	0.0060	1.333	transmembrane protein 59-like	Membrane protein
GgaAffx.20738.1.S1_s_at	TPM1*	0.0290	1.330	tropomyosin 1 (alpha)	Cytoskeleton organization
Gga.5369.1.S1_at	SCNN1G	0.0311	1.330	sodium channel, nonvoltage-gated 1, gamma	Sodium ion transport
Gga.7008.1.S1_at	ALDH4A1*	0.0407	1.328	aldehyde dehydrogenase 4 family, member A1	Proline catabolic process
Gga.5987.1.S1_at	*	0.0455	1.328	Finished cDNA, clone ChEST1003j19	Unknown
Gga.1140.1.S1_at	ERAP1*	0.0193	1.321	endoplasmic reticulum aminopeptidase 1	Proteolysis
Gga.2794.1.S1_at	EP300*	0.0287	1.321	E1A binding protein p300	Signal transduction
Gga.746.1.S1_at	PNAT10*	0.0462	1.321	N-acetyltransferase, pineal gland isozyme NAT-10	Acetyltransferas
Gga.2900.1.S1_at	LOC417353*	0.0428	1.321	hypothetical gene supported by BX933423	Unknown
GgaAffx.12686.1.S1_at	LOC421845*	0.0219	1.314	similar to C6orf37	Unknown
Gga.5588.1.S1_at	PDZK1IP1	0.0347	1.314	PDZK1 interacting protein 1	Signal transduction

Gga.17784.1.S1_at	*	0.0422	1.312	Finished cDNA, clone ChEST887j4	Unknown
Gga.730.1.S1_at	GOT1*	0.0486	1.311	glutamic-oxaloacetic transaminase 1, soluble	Cellular response to insulin stimulus
Gga.5969.2.S1_a_at	PAQR7	0.0111	1.305	progestin and adipoQ receptor family member VII	Receptor activity
Gga.15889.1.S1_at	PCMT1	0.0105	1.304	protein-L-isoaspartate O-methyltransferase	Protein repair
GgaAffx.26355.1.S1_at	SLC10A4	0.0343	1.304	solute carrier family 10, member 4	Sodium/bile acid cotransporter family
GgaAffx.13081.1.S1_s_at	SNAPC5*	0.0136	1.302	small nuclear RNA activating complex, 5	Regulation of transcription
Gga.13758.1.S1_at	*	0.0153	1.302	Finished cDNA, clone ChEST667h20	Unknown
Gga.14250.1.S1_at		0.0449	1.302	Finished cDNA, clone ChEST435d11	Unknown
GgaAffx.24202.1.S1_at	RANBP10	0.0263	1.299	RAN binding protein 10	Signal transduction
Gga.7820.1.S1_at	RIC8B	0.0229	1.297	Resistance to inhibitors of cholinesterase 8B	Signaling
Gga.4569.1.S1_at	PIGR*	0.0205	1.295	polymeric immunoglobulin receptor	Transport immunglobulins
GgaAffx.11201.3.S1_at	LOC430629	0.0458	1.294	hypothetical LOC430629	Unknown
Gga.16371.1.S1_s_at	STT3A*	0.0407	1.287	subunit of oligosaccharyltransferase complex, A	Protein amino acid glycosylation
GgaAffx.8002.1.S1_at	LOC418064*	0.0257	1.285	similar to DNA, Chr 10, Wayne State University 52	Unknown
Gga.1923.1.S1_a_at	CRYZL1	0.0369	1.284	crystallin, zeta (quinone reductase)-like 1	NADPH:quinone reductase activity
Gga.3310.1.S1_at	XPA; XPAC*	0.0457	1.279	xeroderma pigmentosum, complementation A	Nucleotide-excision repair
GgaAffx.12338.2.A1_at	LOC416908	0.0379	1.279	similar to SRR1-like protein	Unknown
GgaAffx.20287.1.S1_at		0.0441	1.274	Finished cDNA, clone ChEST817c22	Unknown
Gga.8796.2.S1_s_at	COPG*	0.0417	1.267	coatomer protein complex, subunit gamma	COPI coating of Golgi vesicle
Gga.18113.1.S1_at	FOXRED1*	0.0309	1.264	FAD-dependent oxidoreductase domain cont. 1	Oxidation reduction
GgaAffx.12236.1.S1_at	UCHL5	0.0452	1.264	ubiquitin carboxyl-terminal hydrolase L5	Ubiquitin cycle
Gga.2826.1.S1_at	*	0.0343	1.263	Finished cDNA, clone ChEST735g14	Unknown
GgaAffx.2936.1.S1_at	SLC7A10	0.0310	1.263	solute carrier family 7, (, y+ system) member 10	Neutral amino acid transporter
Gga.17038.1.S1_at		0.0037	1.261	Finished cDNA, clone ChEST770g13	Unknown
GgaAffx.21799.1.S1_s_at	CMTM3*	0.0450	1.259	CKLF-like MARVEL transmembrane domain 3	Chemotaxis
GgaAffx.12311.1.S1_at	EDEM1*	0.0297	1.258	ER degradation enhancer, mannosidase α-like 1	Response to unfolded protein
GgaAffx.12922.1.S1_at	GFPT1*	0.0445	1.258	glutamine-fructose-6-phosphate transaminase 1	Fructose 6-phosphate metabolic
Gga.2680.1.S1_at	AQP1	0.0465	1.258	aquaporin 1	Water transport
Gga.7684.1.S1_at	MRPL46*	0.0285	1.256	mitochondrial ribosomal protein L46	Ribosomal protein
GgaAffx.3582.1.S1_s_at	SQRDL*	0.0223	1.248	sulfide quinone reductase-like (yeast)	Oxidoreductase activity
Gga.8848.1.S1_at		0.0210	1.248	Finished cDNA, clone ChEST798d14	Unknown
Gga.10831.1.S1_at	HGD*	0.0119	1.247	homogentisate 1,2-dioxygenase	L-phenylalanine / tyrosine catabolism
Gga.10461.1.S1_at	HMGCLL1	0.0198	1.239	3-hydroxymethyl-3-methylglutaryl-CoA lyase-like 1	Lyase activity
Gga.4600.1.S1_at	*	0.0302	1.238	Finished cDNA, clone ChEST1014k22	Unknown
Gga.4060.1.S1_at	PRDX4*	0.0322	1.236	peroxiredoxin 4	I-kappaB phosphorylation
Gga.5561.1.S1_at	RIF1	0.0009	1.235	RAP1 interacting factor homolog (yeast)	Response to DNA damage stimulus
Gga.5464.1.S1_at	ACTR10	0.0327	1.233	actin-related protein 10 homolog (S. cerevisiae)	Microtubule-based movement
GgaAffx.6199.2.S1_s_at	LOC421384	0.0414	1.221	similar to hypothetical protein MGC29875	Unknown
GgaAffx.24365.1.A1_s_at	NUDCD3	0.0305	1.221	NudC domain containing 3	Cell cycle
AFFX-r2-Ec-bioD-3_at	bioD*	0.0201	1.220	dethiobiotin synthetase	Unknown
Gga.490.1.S1_at	CHRD	0.0425	1.220	chordin	Central nervous system development
Gga.7711.2.S1_s_at	WDR57	0.0110	1.218	WD repeat domain 57 (U5 snRNP specific)	RNA splicing
GgaAffx.12351.1.S1_s_at	ING3*	0.0371	1.214	inhibitor of growth family, member 3	Regulation of transcription
GgaAffx.12289.1.S1_at	FAM102A*	0.0369	1.212	family with sequence similarity 102, member A	Estrogen response gene
Gga.3096.1.S1_at	ZNF828*	0.0439	1.205	zinc finger protein 828	Metal ion binding
GgaAffx.555.1.S1_s_at	PRKRIR	0.0313	1.192	interferon-inducible ds RNA dependent inhibitor	Response to stress
Upregulated in DON 2.5					
Gga.15987.1.S1_at	*	0.0116	2.208	Finished cDNA, clone ChEST481o6	Unknown
Gga.554.1.S1_at	HGF	0.0033	2.105	hepatocyte growth factor	Growth factor activity
Gga.16816.1.S1_at	*	0.0314	2.053	Finished cDNA, clone ChEST734i19	Unknown
GgaAffx.20483.1.S1_at		0.0298	1.976	Finished cDNA, clone ChEST655p17	Unknown
GgaAffx.8024.1.S1_at		0.0015	1.887	Transcribed locus	Unknown
GgaAffx.9733.1.S1_at	LOC421765	0.0069	1.855	similar to hypothetical protein FLJ37396	Unknown
Gga.7604.2.S1_a_at	PLSCR1	0.0116	1.812	phospholipid scramblase 1	Phospholipid scrambling
Gga.5517.1.S1_s_at	LOC421267	0.0306	1.799	similar to nuclear DNA-binding protein	Apoptosis inducing
GgaAffx.24639.1.S1_at	FILIP1L	0.0353	1.764	filamin A interacting protein 1-like	Cell proliferation
Gga.4578.1.A1_at		0.0110	1.724	Transcribed locus	Unknown
Gga.19487.1.S1_s_at		0.0315	1.689	Clone cDNA37F microsatellite MCW110 sequence	Unknown
Gga.3674.1.S1_at	STK39	0.0007	1.650	serine threonine kinase 39	Response to cellular stress
Gga.6223.1.S1_at	ADAM33	0.0140	1.621	ADAM metallopeptidase domain 33	Proteolysis
GgaAffx.3622.2.S1_s_at	ABCG2*	0.0308	1.600	ATP-binding cassette, sub-family G, member 2	Response to drug biomarker
GgaAffx.7445.2.S1_s_at	CSTF3	0.0293	1.560	cleavage stimulation factor, 3' pre-RNA, subunit 3	mRNA processing
Gga.16921.1.S1_at	TCP11L2*	0.0416	1.558	t-complex 11 (mouse)-like 2	Multicellular organismal development

Gga.2439.1.S1_at	TMED10*	0.0176	1.511	transmembrane emp24-like trafficking protein 10	Vesicle mediated transport
GgaAffx.26112.1.S1_at	FUCA1	0.0159	1.511	fucosidase, alpha-L- 1, tissue	N-Glycan degradation
Gga.15110.1.S1_at	SFXN3	0.0142	1.488	Sideroflexin 3	Transmembrane transport
Gga.17337.1.S1_at		0.0235	1.481	Finished cDNA, clone ChEST574a9	Unknown
Gga.4193.1.S1_at		0.0398	1.462	Finished cDNA, clone ChEST59b19	Unknown
Gga.5315.1.S1_a_at	LOC417215	0.0329	1.447	similar to RIKEN cDNA 1110008P14	Unknown
Gga.18678.1.S1_at	LOC426456	0.0325	1.439	hypothetical gene supported by CR391215	Unknown
GgaAffx.6854.2.S1_s_at	C8A*	0.0156	1.431	complement component 8, alpha polypeptide	Complement activation
GgaAffx.10750.1.S1_at	MFSD9	0.0328	1.429	major facilitator superfamily domain containing 9	Tetracycline transport
Gga.16717.1.S1_at	*	0.0389	1.414	Finished cDNA, clone ChEST816p14	Unknown
GgaAffx.12703.1.S1_at	LOC415623	0.0197	1.410	similar to Toll-like receptor 21	Innate immune response
GgaAffx.5975.1.S1_at	RAD51L1	0.0276	1.395	RAD51-like 1 (S. cerevisiae)	DNA repair
GgaAffx.24006.1.A1_at	LOC420882	0.0232	1.387	similar to chromosome 6 open reading frame 86	Unknown
GgaAffx.21414.1.S1_s_at	INTS6*	0.0379	1.381	integrator complex subunit 6	snRNA processing
Gga.10286.1.S1_at	PBLD*	0.0150	1.351	phenazine biosynthesis-like protein dom. cont.	Biosynthetic process
Gga.15844.1.S1_s_at	*	0.0204	1.351	Hypothetical protein, clone 14a17	Unknown
GgaAffx.12611.1.S1_at	ANGEL1	0.0147	1.346	angel homolog 1 (Drosophila)	CCR4/nocturin family
GgaAffx.11878.1.S1_s_at	SFRS5	0.0080	1.342	splicing factor, arginine/serine-rich 5	mRNA processing
GgaAffx.11907.1.S1_s_at	DDX27	0.0321	1.339	DEAD (Asp-Glu-Ala-Asp) box polypeptide 27	RNA helicase
Gga.12914.1.S1_at	LOC418213	0.0251	1.337	similar to Hypothetical protein FLJ10637	Unknown
Gga.5359.1.S1_at	LOC417676	0.0165	1.326	hypothetical LOC417676	Unknown
Gga.13587.1.S1_s_at	USP38	0.0434	1.325	ubiquitin specific peptidase 38	Ubiquitin cycle
Gga.6387.1.S1_at	RAC2*	0.0038	1.321	ras-related C3 botulinum toxin substrate 2	Signal transduction
Gga.16161.1.S1_at	SLC25A33	0.0300	1.318	solute carrier family 25, member 33	Mitochondrial transporter
AFFX-r2-Ec-bioB-3_at	bioB	0.0152	1.316	Biotin synthase ; biotin synthesis, sulfur insertion	Biotin synthesis
Gga.12238.2.S1_a_at	ATP6V0E2	0.0125	1.311	ATPase, H+ transporting V0 subunit e2	Unknown
GgaAffx.12745.1.S1_at	SMYD5	0.0283	1.307	SMYD family member 5	Retinoic acid induced
Gga.4664.2.S1_a_at		0.0101	1.302	Finished cDNA, clone ChEST252c24	Unknown
Gga.20060.1.S1_at	ABAT*	0.0191	1.287	4-aminobutyrate aminotransferase	γ-aminobutyric acid catabolic process
Gga.20020.1.S1_at	PKNOX2	0.0423	1.285	PBX/knotted 1 homeobox 2	Transcription regulation
Gga.1520.1.S1_at	*	0.0109	1.282	Finished cDNA, clone ChEST164h12	Unknown
Gga.13287.2.S1_s_at	UBE2D1	0.0411	1.261	ubiquitin-conjugating enzyme E2D 1	E2 ubiquitin cycle
GgaAffx.21492.1.S1_at	*	0.0175	1.259	Finished cDNA, clone ChEST1037o8	Unknown
GgaAffx.20443.1.S1_at	LOC418032	0.0468	1.256	similar to mKIAA1668 protein	Unknown
Gga.4890.2.S1_a_at	SMAP1L	0.0344	1.252	stromal membrane-associated protein 1-like	Vesicle trafficking
Gga.9991.1.S1_at	SORD*	0.0295	1.248	sorbitol dehydrogenase	Fructose / mannose metabolism
GgaAffx.22408.1.S1_at	LOC429596	0.0115	1.242	similar to keratin	Cytoskeleton
Gga.19067.1.S1_at	RAB7A*	0.0032	1.222	RAB7A, member RAS oncogene family	Endocytosis
Gga.7026.1.S1_at	LOC425727*	0.0131	1.218	similar to natterin 3	Unknown
GgaAffx.12757.1.S1_s_at	RGL1	0.0367	1.206	ral guanine nucleotide dissociation stimulator-like 1	Signal transduction
Gga.7034.1.S1_at	*	0.0326	1.202	similar to dullard homolog [Canis familiaris]	Nuclear membrane organization
GgaAffx.13069.2.S1_s_at	NFRKB*	0.0293	1.202	nuclear factor related to kappaB binding protein	Inflammatory response
Gga.13246.1.S1_at		0.0047	1.202	Finished cDNA, clone ChEST405a9	Unknown

Upregulated in DON 1

Gga.5787.1.S1_at		0.0183	4.255	Transcribed locus	Unknown
Gga.133.1.S1_at	LOC422661*	0.0094	2.557	similar to sulfotransferase 1B	Xenobiotic metabolism
Gga.18298.1.S1_at	RAPGEF2	0.0396	2.336	Rap guanine nucleotide exchange factor (GEF) 2	MAPKKK
Gga.13112.1.S1_at		0.0358	2.283	Finished cDNA, clone ChEST509f16	Unknown
Gga.13301.1.S1_at	GDPD5*	0.0107	2.208	glycerophosphodiester phosphodiesterase dom. 5	Glycerol metabolism
Gga.14825.1.S1_at		0.0392	2.169	Finished cDNA, clone ChEST39c19	Unknown
Gga.8138.1.S1_at	ELF3	0.0240	2.119	E74-like factor 3	Inflamatorry response
GgaAffx.24837.1.S1_at	KLF10*	0.0191	2.070	Kruppel-like factor 10	Transcription factor
Gga.14716.1.S1_at		0.0412	1.961	Finished cDNA, clone ChEST498b13	Unknown
Gga.15973.1.S1_at	RIPK4*	0.0070	1.919	receptor-interacting serine-threonine kinase 4	Protein amino acid phosphorylation
GgaAffx.20820.1.S1_s_at	SNAP23*	0.0253	1.883	Synaptosomal-associated protein, 23kDa	Cellular membrane fusion
Gga.7554.1.S1_s_at	RAB22A	0.0115	1.883	RAB22A, member RAS oncogene family	Endosome organization and
Gga.16391.1.S1_at		0.0269	1.855	Finished cDNA, clone ChEST620d2	Unknown
Gga.6088.1.S1_a_at	NDUFA7*	0.0304	1.841	NADH dehydrogenase 1 α subcomplex, 7, 14.5kDa	Mitochondrial electron transport, NADH
Gga.5312.1.S1_at	SNX22	0.0483	1.825	sorting nexin 22	Protein transport
Gga.18324.1.S1_s_at		0.0151	1.818	Finished cDNA, clone ChEST228a16	Unknown
GgaAffx.11973.1.S1_s_at	LOC770547*	0.0180	1.805	Similar to pancreatic intrinsic factor	Cobalamin transport
Gga.8078.1.S1_at	GAS6	0.0109	1.792	growth arrest-specific 6	Cell proliferation
Gga.8487.2.S1_at	SRP72	0.0194	1.789	signal recognition particle 72kDa	Signal transducer

GgaAffx.8027.1.S1_at	LOC420821	0.0339	1.761	similar to Doublecortin domain-containing protein 2	Cellular defense response
Gga.224.1.S1_at	ZP1	0.0186	1.745	zona pellucida glycoprotein 1 (sperm receptor)	Glycoprotein
Gga.7782.1.S1_at		0.0091	1.742	Finished cDNA, clone ChEST250a3	Unknown
Gga.7894.1.S1_at	SEMA5A*	0.0268	1.727	sema domain, 7 transmembrane dom., 5A	Cell adhesion
Gga.7001.1.S1_at	*	0.0229	1.712	Finished cDNA, clone ChEST560d24	Unknown
Gga.19396.1.S1_s_at	RAB22A	0.0021	1.669	RAB22A, member RAS oncogene family	Endosome organization / biogenesis
Gga.17757.1.S1_at		0.0388	1.667	Finished cDNA, clone ChEST790f21	Unknown
GgaAffx.24772.1.S1_at	LOC421818	0.0337	1.639	hypothetical LOC421818	Unknown
Gga.11828.1.S1_at	NINJ1	0.0155	1.631	ninjurin 1	Induction senescence
Gga.6153.1.S1_s_at		0.0097	1.603	Hypothetical protein, clone 9e23	Unknown
GgaAffx.12460.1.S1_s_at	NPC2*	0.0196	1.600	Niemann-Pick disease, type C2	Lipid transport
Gga.895.1.S1_at	S100A6	0.0085	1.585	S100 calcium binding protein A6	Signal transducer
Gga.9810.1.S1_s_at	GMPPB*	0.0184	1.531	GDP-mannose pyrophosphorylase B	Fructose / mannose metabolism
Gga.983.1.S1_at		0.0318	1.513	Finished cDNA, clone ChEST916c16	Unknown
Gga.791.1.S1_at	IRF1*	0.0493	1.497	interferon regulatory factor 1	Transcription factor
Gga.5950.1.S1_s_at		0.0340	1.493	Finished cDNA, clone ChEST872m23	Unknown
Gga.7272.1.S1_at	*	0.0345	1.484	Finished cDNA, clone ChEST626b13	Unknown
Gga.5159.1.S1_at	MYST3*	0.0094	1.402	MYST histone acetyltransferase 3	Histone acetylation
Gga.3802.2.S1_a_at	PDIA3*	0.0277	1.401	protein disulfide isomerase family A, member 3	Protein import in nucleus ER
Gga.11286.1.S1_at		0.0450	1.401	Finished cDNA, clone ChEST742e16	Unknown
Gga.1210.1.S2_at	CCRN4L	0.0268	1.399	CCR4 carbon catabolite repression 4-like)	Transcript. RNA polymerase II promo.
GgaAffx.2053.1.S1_at	LOC425118	0.0335	1.387	similar to R31449_3	Unknown
Gga.19451.1.S1_at	ADAM28	0.0047	1.385	ADAM metallopeptidase domain 28	Cell proliferation
Gga.9513.1.S2_at	CXCL12	0.0242	1.383	chemokine (C-X-C motif) ligand 12	Immune response
Gga.5171.1.S1_at	CDH5*	0.0385	1.359	cadherin 5, type 2, VE-cadherin	Cell adhesion
Gga.4896.1.S1_at	LOC427638*	0.0449	1.357	similar to glutathione peroxidase (EC 1.11.1.9)	Anti-Apoptosis
Gga.14007.1.S1_at		0.0347	1.299	Finished cDNA, clone ChEST886a8	Unknown
Gga.6941.1.S1_at		0.0397	1.299	Finished cDNA, clone ChEST594i1	Unknown
Gga.13415.1.S1_at	ADD3*	0.0046	1.287	adducin 3 (gamma)	Cytoskeletal protein

*: Raw value above the 80th percentile.

Table 8.2: 1st Deoxynivalenol experiment; downregulated genes

Probe Set ID	Name	P-value	Fc	Description	Function
Downregulated in DON 1, DON 2.5 and DON 5					
Gga.19878.1.S1_at	ARID1A	0.0059	1.664	AT rich interactive domain 1A (SWI-like)	Androgen/estrogen receptor sig.
GgaAffx.6720.3.S1_s_at	ZC3H14	0.0055	1.332	zinc finger CCCH-type containing 14	Nucleic acid binding
GgaAffx.26265.2.S1_s_at	ATXN2	0.0055	1.315	ataxin 2	RNA metabolic process
Gga.14829.1.S1_at	LOC423360	0.0055	1.249	similar to DNA mismatch repair protein Mlh3	DNA repair
Downregulated in DON 2.5 and DON 5					
GgaAffx.26.1.S1_at	LOC427047	0.0209	2.089	similar to breast cancer-associated antigen BRCAA1	Transcription regulation
GgaAffx.12182.1.S1_at	ODF2	0.0092	1.701	outer dense fiber of sperm tails 2	Cell differentiation
GgaAffx.5339.1.S1_at	EXOC4	0.0050	1.676	exocyst complex component 4	Protein transport
GgaAffx.25933.3.S1_s_at	PITPNC1	0.0058	1.592	phosphatidylinositol transfer protein, cytoplasmic 1	Lipid transport
Gga.15747.2.S1_s_at	APRIN	0.0174	1.535	androgen-induced proliferation inhibitor	Cell cycle
GgaAffx.20644.1.S1_s_at	FAM107B*	0.0180	1.489	Family with sequence similarity 107, member B	Unknown
GgaAffx.26385.1.S1_s_at		0.0003	1.445	Finished cDNA, clone ChEST621g22	Unknown
Gga.14100.1.S1_s_at	HHIP	0.0026	1.369	hedgehog interacting protein	Signal transduction
Gga.15723.1.S1_s_at	LOC416654	0.0054	1.369	hypothetical LOC416654	Unknown
GgaAffx.20535.1.S1_at		0.0181	1.369	Finished cDNA, clone ChEST1015a13	Unknown
Gga.12970.1.S1_at		0.0157	1.359	Finished cDNA, clone ChEST388i12	Unknown
Gga.16696.1.S1_at	PPP2R5E	0.0083	1.349	protein phosphatase 2, regulatory subunit B', epsilon	Signal transduction
GgaAffx.2569.1.S1_at	USP20	0.0291	1.32	ubiquitin specific peptidase 20	Protein deubiquitination
Gga.6043.2.S1_at		0.0139	1.317	similar to PTTG1 protein-interacting protein	Protein import into nucleus
GgaAffx.24078.1.S1_s_at	TRIO	0.0068	1.313	triple functional domain (PTPRF interacting)	Signal transduction
GgaAffx.20844.1.S1_s_at	LOC422491	0.0272	1.302	similar to hypothetical protein FLJ23356	Unknown
Gga.9210.1.S1_at		0.0060	1.27	Transcribed locus	Unknown
GgaAffx.9982.4.S1_s_at	ADD1	0.0039	1.213	adducin 1 (alpha)	Actin binding
Downregulated in DON 1 and DON 5					
Gga.9154.3.S1_x_at	PACS2	0.0383	1.712	phosphofurin acidic cluster sorting protein 2	Apoptosis

Gga.7044.3.S1_at	SEPP1	0.0199	1.426	selenoprotein P, plasma, 1	Response to oxidative stress
GgaAffx.7282.1.S1_at	GPD1L	0.0099	1.247	glycerol-3-phosphate dehydrogenase 1-like	Carbohydrate metabolic process
Downregulated in DON 1 and DON 2.5					
Gga.9838.1.S1_at		0.0003	3.107	Finished cDNA, clone ChEST956f13	Unknown
Gga.6220.5.S1_a_at	ICER	0.0063	2.029	ICER protein	Transcription regulation
Gga.2445.1.S1_at	LOC422442	0.0188	1.562	hypothetical gene supported by CR385915	Unknown
Gga.14677.1.S1_at		0.0274	1.508	Finished cDNA, clone ChEST780f19	Unknown
GgaAffx.25502.2.S1_at	SLC2A5*	0.0043	1.259	solute carrier family 2, member 5	Glucose / fructose transport
Gga.4703.1.S1_s_at	NDFIP1*	0.0305	1.2000	Nedd4 family interacting protein 1	I-κB kinase/NF-κB regulation
Downregulated in DON 5					
Gga.9580.2.S1_a_at	AKR1B1*	0.0468	3.948	aldo-keto reductase family 1, member B1	Response to stress
Gga.15340.1.S1_at	*	0.0086	2.991	Finished cDNA, clone ChEST581b17	Unknown
GgaAffx.11878.1.S1_s_at	MIA2	0.0126	2.544	melanoma inhibitory activity 2	Inflammation
GgaAffx.11667.1.S1_s_at	LOC416628*	0.0356	2.261	similar to hypothetical protein MGC24665	DNA binding domain
GgaAffx.6850.1.S1_at	BTBD7	0.0010	1.769	BTB (POZ) domain containing 7	Related to tumorigenesis
Gga.4930.3.S1_a_at	LOC424773	0.0150	1.693	similar to KIAA0332	RNA processing
GgaAffx.12172.1.S1_s_at	PREPL	0.0454	1.686	prolyl endopeptidase-like	Proteolysis
Gga.14891.1.S1_s_at	RBBP6	0.0388	1.651	retinoblastoma binding protein 6	Protein ubiquitination
Gga.9934.1.S1_s_at	ANKRD13A	0.0371	1.631	ankyrin repeat domain 13A	DNA damage related
Gga.2310.1.S1_at	LMO4	0.0340	1.629	LIM domain only 4	Transcription regulation
Gga.10986.1.S1_at	*	0.0144	1.608	Finished cDNA, clone ChEST652g13	Unknown
GgaAffx.20476.1.S1_at		0.0082	1.593	Finished cDNA, clone ChEST653j6	Unknown
Gga.13135.2.S1_a_at	ZCRB1*	0.0234	1.586	Zinc finger CCHC-type and RNA binding motif 1	RNA splicing
Gga.9603.1.S1_s_at	USP8	0.0207	1.529	ubiquitin specific peptidase 8	Cell proliferation
Gga.7868.1.S1_at	*	0.0327	1.525	Finished cDNA, clone ChEST737n20	Unknown
Gga.12626.2.S1_s_at	HCFC2	0.0027	1.524	host cell factor C2	Transcription regulation
GgaAffx.26072.2.S1_s_at	TJP1	0.0248	1.52	tight junction protein 1 (zona occludens 1)	Intercellular junction assembly
Gga.3068.1.S1_at	LOC419622*	0.0279	1.511	similar to TRAP150	Androgen receptor signaling
Gga.2103.3.S1_a_at		0.0162	1.509	Finished cDNA, clone ChEST935l6	Unknown
GgaAffx.20095.1.S1_at	EXOSC9	0.0099	1.503	exosome component 9	Immune response
Gga.8418.1.S1_at		0.0230	1.503	Finished cDNA, clone ChEST1023m18	Unknown
GgaAffx.26106.6.S1_s_at	LOC428354	0.0359	1.48	similar to Jumonji domain containing protein 2B	Transcription regulation
Gga.16149.1.S1_s_at	ANKMY1	0.0237	1.474	ankyrin repeat and MYND domain containing 1	Protein binding
GgaAffx.7493.1.S1_s_at	LOC422658*	0.0371	1.473	similar to UDP-glucuronosyltransferase 2A1 precursor	Detection of chemical stimulus
GgaAffx.8082.2.A1_at	ELOVL2*	0.0301	1.472	elongation of very long chain fatty acids -like 2	Fatty acid biosynthesis
Gga.2633.1.S1_at	FBXL14; PPA	0.0462	1.469	F-box and leucine-rich repeat protein 14	Ubiquitin cycle
GgaAffx.21354.1.S1_at	YPEL1*	0.0390	1.438	yippee-like 1 (Drosophila)	Cell morphology
GgaAffx.13221.1.S1_s_at	MTPN	0.0031	1.425	Myotrophin	Cell growth
Gga.1752.1.S1_at		0.0357	1.419	Finished cDNA, clone ChEST654l15	Unknown
GgaAffx.8767.1.S1_at	LOC421667	0.0075	1.409	similar to Phosphatase and actin regulator 2	Actin binding
GgaAffx.25504.1.S1_at	GPR157	0.0217	1.404	G protein-coupled receptor 157	Signal transduction
Gga.3397.1.S1_at	NAT2	0.0444	1.401	N-acetyltransferase 2 (arylamine N-acetyltransferase)	Drug metabolism
Gga.11668.1.S1_at	NDRG1*	0.0287	1.395	N-myc downstream regulated gene 1	Response to metal ion
Gga.9939.1.S1_a_at	DTD1; HARS2	0.0310	1.394	D-tyrosyl-tRNA deacylase 1 homolog (S. cerevisiae)	D-amino acid catabolic process
Gga.2569.3.S1_a_at	TERF2; TRF2	0.0282	1.392	telomeric repeat binding factor 2	Telomere maintainance
Gga.12905.1.S1_at	MARK1*	0.0324	1.391	MAP/microtubule affinity-regulating kinase 1	Protein kinase cascade
GgaAffx.20792.1.S1_s_at	ZEB1; TCF8	0.0295	1.37	zinc finger E-box binding homeobox 1	Neg. reg. of transcription
GgaAffx.24534.1.S1_at	LOC427188	0.0376	1.366	similar to Complement component C7 precursor	Complement activation
Gga.5598.1.S1_at		0.0060	1.362	Transcribed locus	Unknown
GgaAffx.5682.1.S1_at	SRI	0.0008	1.352	Sorcin	Signal transduction
Gga.4564.2.S1_a_at	FMO6; FMO3*	0.0238	1.345	flavin containing monooxygenase 6	Transport
Gga.7965.1.S1_at	RAMP3	0.0071	1.343	receptor (G protein-coupled) activity modifying prot. 3	Receptor-mediated endocytosis
Gga.212.1.S1_at	NRXN1*	0.0357	1.342	neurexin 1	Receptor-mediated endocytosis
Gga.8268.1.S1_s_at	NDST2	0.0426	1.339	N-deacetylase/N-sulfotransferase 2	Heparan sulfate biosynthesis
Gga.10243.1.S1_at	C8B*	0.0094	1.336	complement component 8, beta polypeptide	Complement activation
GgaAffx.20662.1.S1_at	NARF	0.0267	1.336	nuclear prelamin A recognition factor	Lamin binding
GgaAffx.26456.1.S1_s_at	FGD3*	0.0242	1.333	FYVE, RhoGEF and PH domain containing 3	Regulation actin cytoskeleton
GgaAffx.20462.1.S1_at		0.0393	1.329	Finished cDNA, clone ChEST766a17	Unknown
Gga.19064.1.S1_at		0.0274	1.324	Finished cDNA, clone ChEST590e8	Unknown
GgaAffx.5091.1.S1_at	LOC427847	0.0196	1.314	similar to KIAA1218 protein	Unknown
GgaAffx.3165.1.S1_at	LOC422195	0.0447	1.309	Methyltransferase activity	Unknown
Gga.13529.1.S1_s_at	SLC31A1*	0.0193	1.305	Solute carrier family 31,member 1	Copper transport

Gga.9724.1.S1_at	STEAP3	0.0132	1.303	STEAP family member 3	Apoptosis
Gga.9740.1.S1_at	PPCDC	0.0363	1.301	phosphopantothenoylcysteine decarboxylase	Coenzyme A biosynthesis
Gga.4564.1.S1_a_at	FMO6; FMO3*	0.0296	1.293	flavin containing monooxygenase 6	Transport
Gga.17342.1.S1_at		0.0482	1.291	Finished cDNA, clone ChEST845e14	Unknown
Gga.7689.1.S1_at	CASK*	0.0063	1.284	calcium/calmodulin-dependent serine protein kinase	Cell adhesion
GgaAffx.6314.1.S1_s_at	RBKS	0.0022	1.284	Ribokinase	Pentose phosphate pathway
GgaAffx.2219.3.S1_s_at	PEX11G	0.0375	1.276	peroxisomal biogenesis factor 11 gamma	Peroxisome
Gga.6922.1.S1_at	MYSM1	0.0352	1.273	myb-like, SWIRM and MPN domains 1	Transcription regulation
GgaAffx.25682.1.S1_s_at	NEDD4L	0.0128	1.268	neural precursor cell expressed	Ubiquitination
Gga.2572.1.S1_at	LAMP2	0.0328	1.267	lysosomal-associated membrane protein 2	Glycogen metabolic process
Gga.13294.1.S1_at	*	0.0196	1.245	Finished cDNA, clone ChEST387p23	Unknown
GgaAffx.25215.2.S1_s_at	PCF11	0.0452	1.245	cleavage and polyadenylation factor subunit, homolog	mRNA cleavage
GgaAffx.12803.1.S1_s_at	LARP7	0.0212	1.244	La ribonucleoprotein domain family, member 7	RNA processing
GgaAffx.20765.1.S1_at	SLC45A4	0.0106	1.242	Solute carrier family 45, member 4	Transmembran transport
GgaAffx.20668.1.S1_s_at	ATXN7L1*	0.0200	1.238	Ataxin 7-like 1	Unknown
Gga.16368.1.S1_at		0.0319	1.235	Finished cDNA, clone ChEST239f8	Unknown
Gga.13677.1.S1_at	*	0.0048	1.233	Finished cDNA, clone ChEST1025i5	Unknown
Gga.4558.2.S1_s_at	MXD4	0.0149	1.23	MAX dimerization protein 4	Neg. reg. of transcription
GgaAffx.6980.1.S1_at	RAVER2	0.0441	1.222	ribonucleoprotein, PTB-binding 2	RNA binding
Gga.9798.1.S1_s_at	HSA277841	0.0268	1.22	ELG protein	Unknown
GgaAffx.12896.1.S1_at	SLC7A5	0.0025	1.218	solute carrier family 7 (y+ system), member 5	Cationic amino acid transporter
Gga.1827.1.S1_s_at	*	0.0134	1.207	Finished cDNA, clone ChEST1009j1	Unknown
GgaAffx.20257.1.S1_s_at	SLC36A1	0.0148	1.207	Solute carrier family 36, member 1	Proton/amino acid symporter
GgaAffx.4477.1.S1_at	BRWD3	0.0230	1.205	bromodomain and WD repeat domain containing 3	DNA damage related
Gga.10215.1.S1_at		0.0373	1.204	Finished cDNA, clone ChEST32e24	Unknown
GgaAffx.21693.1.S1_s_at		0.0018	1.204	Finished cDNA, clone ChEST738p7	Unknown
GgaAffx.12606.1.S1_s_at	JAM2	0.0280	1.193	junctional adhesion molecule 2	Cell-cell adhesion
Downregulated in DON 2.5					
Gga.178.1.S1_s_at	GBP	0.0110	2.103	guanylate binding protein	Immune response
Gga.12243.1.S1_at	RWDD2B	0.0226	1.891	RWD domain containing 2B	Unknown
Gga.10529.1.S1_at	LOC770294	0.0078	1.702	Similar to RNase H, putative	DNA integration
Gga.6038.1.S1_at	LOC417856	0.0451	1.649	similar to TMEM19 protein	Multi-pass membrane protein
Gga.11191.1.S1_at	LOC415382	0.0134	1.624	similar to BB128963 protein	Unknown
Gga.1404.2.S1_at	MCFD2*	0.0178	1.605	multiple coagulation factor deficiency 2	Calcium ion binding
Gga.8763.1.S1_at	YTHDC1	0.0249	1.565	YTH domain containing 1	RNA splicing
GgaAffx.21020.1.S1_s_at	LOC427819	0.0345	1.541	similar to bromodom. adjacent to zinc finger dom.1B	Chromatin remodeling
Gga.15593.1.A1_at		0.0160	1.525	Transcribed locus	Transcribed locus
Gga.19440.1.S1_at	LOC423771	0.0096	1.498	Similar to ubiquitin-like fusion protein	Unknown
GgaAffx.21104.1.S1_at	EXOC4*	0.0218	1.489	Exocyst complex component 4	Regulation of protein transport
GgaAffx.13074.1.S1_at	DNAJC3	0.0403	1.435	DnaJ (Hsp40) homolog, subfamily C, member 3	Defense response
GgaAffx.2216.2.S1_s_at	LOC427438	0.0276	1.425	hypothetical LOC427438	Unknown
Gga.6035.1.S1_at		0.0462	1.42	Finished cDNA, clone ChEST38o15	Unknown
Gga.4355.1.S1_at	MAP2K1IP1	0.0321	1.417	MAPK kinase 1 interacting protein 1	MAPK pathway
Gga.10975.1.S1_s_at	ARID1B	0.0397	1.407	similar to AT rich interactive domain 1B isoform 1	Maintenance of transcription
Gga.12555.1.S1_s_at	CCDC55	0.0112	1.394	coiled-coil domain containing 55	Unknown
Gga.9827.1.S1_s_at	GOLIM4	0.0392	1.391	golgi integral membrane protein 4	Transport
GgaAffx.2216.1.S1_at	LOC427438	0.0361	1.391	hypothetical LOC427438	Unknown
Gga.9001.1.S1_at	LOC426427	0.0268	1.363	similar to KIAA0556 protein	Unknown
GgaAffx.12282.1.S1_s_at	MTHFD2	0.0207	1.353	methylenetetrahydrofolate dehydrogenase 2	Tetrahydrofolate metabolism
Gga.572.1.S2_at	NR5A2	0.0261	1.351	nuclear receptor subfamily 5, group A, member 2	Regulation of transcription
Gga.17375.1.S1_at		0.0447	1.348	Finished cDNA, clone ChEST709e8	Unknown
Gga.14258.1.S1_at	ZNF280D	0.0350	1.344	zinc finger protein 280D	Transcription regulation
Gga.1491.1.S1_at	RREB1	0.0442	1.344	ras responsive element binding protein 1	Regulation of transcription
Gga.16426.1.S1_at		0.0032	1.339	Finished cDNA, clone ChEST459g7	Unknown
Gga.16888.1.S1_at		0.0432	1.338	Finished cDNA, clone ChEST437d16	Unknown
GgaAffx.12828.1.S1_at	LOC418842	0.0103	1.338	similar to RIKEN cDNA 1200011I18	Unknown
GgaAffx.21769.1.S1_s_at	LPL*	0.0440	1.327	lipoprotein lipase	Phospholipid metabolic process
GgaAffx.25412.1.S1_s_at	DDX59	0.0262	1.325	DEAD (Asp-Glu-Ala-Asp) box polypeptide 59	RNA binding
Gga.16858.2.S1_a_at		0.0207	1.321	Finished cDNA, clone ChEST373c10	Unknown
GgaAffx.21542.1.S1_at	*	0.0362	1.32	Finished cDNA, clone ChEST563l4	Unknown
Gga.11559.2.S1_a_at	PLEKHA1	0.0319	1.311	pleckstrin homology domain containing, A1	Signal transduction
GgaAffx.10528.1.S1_at	ROCK2	0.0094	1.308	Rho-associated, coiled-coil cont. protein kinase 2	Cytokenesis
GgaAffx.1421.4.S1_s_at	DGKQ*	0.0364	1.304	diacylglycerol kinase, theta 110kDa	Intracellular singaling cascade

Gga.16908.1.S1_s_at	USP25	0.0243	1.298	ubiquitin specific peptidase 25	Proteolysis
GgaAffx.10171.1.S1_at	EIF2AK3	0.0100	1.286	eukaryotic translation initiation factor 2-alpha kinase 3	Neg. regulation of translation
GgaAffx.5884.3.S1_s_at	LOC423251	0.0158	1.273	similar to RIKEN cDNA 2610015J01	Unknown
Gga.16422.1.S1_at		0.0188	1.268	Finished cDNA, clone ChEST377n6	Unknown
Gga.14739.1.S1_s_at	UBR1	0.0134	1.256	ubiquitin protein ligase E3 component n-recognin 1	Ubiquitin cycle
GgaAffx.25124.1.S1_s_at	RBM26	0.0269	1.249	RNA binding motif protein 26	RNA binding
Gga.16793.1.S1_s_at	PDS5A	0.0216	1.246	PDS5, regulator of cohesion maintenance, homolog	Cell cycle
Gga.10898.1.S1_s_at	MTIF2*	0.0149	1.238	mitochondrial translational initiation factor 2	Regulation of translation initiation
Gga.17229.1.S1_at	*	0.0325	1.235	Finished cDNA, clone ChEST887n14	Unknown
Gga.16535.1.S1_s_at	MAP2K5	0.0166	1.232	mitogen-activated protein kinase kinase 5	Signal transduction
Gga.7296.3.S1_s_at	IFNGR2	0.0213	1.231	interferon gamma receptor 2	Signal transduction
Gga.12903.1.S1_at	ILKAP	0.0113	1.225	integrin-link. KA serine/threonine phosphate. 2C	Neg. regulation cell cycle
Gga.17289.1.S1_at		0.0349	1.225	Finished cDNA, clone ChEST576g8	Unknown
Gga.14841.1.S1_s_at	LOC421095	0.0175	1.224	similar to TRS85 homolog	Unknown
Gga.7973.1.S1_at		0.0444	1.217	Finished cDNA, clone ChEST69m10	Unknown
GgaAffx.21002.1.S1_s_at	ADAT1	0.0139	1.209	adenosine deaminase, tRNA-specific 1	tRNA processing
GgaAffx.20293.1.S1_s_at		0.0386	1.207	Finished cDNA, clone ChEST817o13	Unknown
GgaAffx.9887.1.S1_s_at	SGK3	0.0395	1.202	serum/glucocorticoid regulated kinase family, 3	Response to stress
Downregulated in DON 1					
Gga.11825.1.S1_s_at	GK*	0.0465	4.313	glycerol kinase	PPAR signaling
GgaAffx.5600.1.S1_at	EHBP1	0.0451	3.452	EH domain binding protein 1	Related to endocytosis
GgaAffx.6623.3.S1_s_at	LOC424618*	0.0424	2.722	similar to cytochrome P450, 4B1	Xenobiotic metabolism
GgaAffx.4115.3.S1_s_at	ATP11C	0.0423	2.423	ATPase, Class VI, type 11C	Phospholipid transport
Gga.17520.1.S1_at		0.0291	2.417	Finished cDNA, clone ChEST802d6	Unknown
GgaAffx.8013.1.S1_at	MTERFD3	0.0273	2.039	MTERF domain containing 3	Transcription regulation
Gga.13412.1.S1_at	OSBPL2	0.0239	1.912	oxysterol binding protein-like 2	Lipid transport
GgaAffx.22014.2.S1_s_at	EHHADH*	0.0217	1.843	enoyl-Coenzyme A	Lipid metabolic process
Gga.11911.2.S1_a_at	ZMYND11	0.0200	1.816	zinc finger, MYND domain containing 11	Cell proliferation
Gga.4210.1.S1_at	MT4*	0.0181	1.774	metallothionein 4	Cellular metal ion homeostasis
Gga.6043.2.S1_a_at	*	0.0177	1.605	similar to PTTG1 protein-interacting protein	Protein transport
GgaAffx.22298.1.S1_s_at	LOC416353*	0.0161	1.449	similar to GFR receptor alpha 4	Receptor activity
Gga.16363.1.S1_at	NKTR	0.0148	1.441	natural killer-tumor recognition sequence	Protein folding
GgaAffx.20100.1.S1_at	CPT1A; CPT1*	0.0148	1.434	carnitine palmitoyltransferase 1A (liver)	Fatty acid beta-oxidation
Gga.7542.1.S1_at	EHHADH*	0.0105	1.416	enoyl-Coenzyme A	Fatty acid metabolic process
GgaAffx.25502.1.S1_s_at	SLC2A5*	0.0013	1.404	solute carrier family 2, member 5	Glucose / fructose transport
Gga.10416.1.S1_s_at	CPT1A	0.0001	1.264	carnitine palmitoyltransferase 1A (liver)	Fatty acid beta-oxidation
GgaAffx.21713.1.S1_at		0.0128	1.422	Finished cDNA, clone ChEST401p6	Unknown

*: Raw value above the 80th percentile.

9. Appendix C

Table 9.1: 2nd Deoxynivalenol experiment; significantly regulated genes in D20

Probe Set ID, Gga.	Other groups	Fc	Gene symbol	Gene title	Gene function
Apoptosis					
3144.1.S1_at	D20 vs MD20	1.392	TM2D1	TM2 domain containing 1	Apoptosis
6233.1.S1_at	D20 vs MD20	-1.266	SON	SON DNA binding protein	Anti-apoptosis
8546.1.S1_a_at		-1.285	LOC378902	death domain-containing TNF receptor, 23	Apoptosis regulation
Affx.12013.1.S1_at		-1.305	CAT	Catalase	Negative regulation of apoptosis
Cell junction					
Affx.4655.1.S1_at		-1.209	TMEM8	transmembrane protein 8	Cell adhesion
3635.2.S1_s_at		-1.215	CTTN	Cortactin	Tight junction regulation
11213.1.S1_s_at	D20 vs MD20	-1.289	CGNL1[2]	cingulin-like 1	Cell junction
Cell cycle, growth proliferation					
4066.1.S1_at	D20 vs MD20	1.488	SMC2	structural maintenance of chromosomes 2	Cell division
Affx.12961.1.S1_at	D20 vs MD20	1.477	TPX2	TPX2, microtubule-associated, homolog	Cell proliferation
18957.1.S1_at	D20 vs MD20	1.438	MKI67	antigen identified by antibody Ki-67	Cell proliferation
8287.1.S1_at		1.435	TIPRL	TIP41, TOR signalling pathway regulator-like	DNA damage checkpoint
11746.1.S1_at	D20 vs MD20	1.414	PBK	PDZ binding kinase	Mitosis
726.2.S1_a_at	D20 vs MD20	1.381	CDC2	cell division cycle 2, G1 to S and G2 to M	Cell cycle
Affx.11578.1.S1_s_at	D20 vs MD20	1.354	KPNA2	karyopherin alpha 2	Cell cycle
421.2.S1_a_at		1.323	TK1	thymidine kinase 1, soluble	DNA replication
2844.2.S1_s_at	D20 vs MD20	1.297	STMN1[2]	stathmin 1/oncoprotein 18	Mitotic spindle organization
1551.2.S1_a_at	D20 vs MD20	1.289	BIRC5	baculoviral IAP repeat-containing 5	Cell division
2010.1.S1_at	D20 vs MD20	1.256	TOP2A	topoisomerase (DNA) II alpha 170kDa	DNA ligation
10003.1.S1_at	D20 vs MD20	-1.205	MKL2	MKL/myocardin-like 2	Cell differentiation
9707.1.S1_s_at		-1.205	FNIP2	folliculin interacting protein 2	Cell cycle
Affx.4473.1.S1_at	MD20	-1.209	FAM107A	family with sequence similarity 107, A	Cell growth regulation
Affx.25078.1.S1_at		-1.222	SEPT10	septin 10	Cell cycle
10852.1.S1_s_at	D20 vs MD20	-1.224	LRRC8A	leucine rich repeat containing 8 family, A	Cell differentiation
Affx.7262.1.S1_at		-1.292	TNFAIP2	tumor necrosis factor, alpha-induced protein 2	Cell differentiation
1331.2.S1_a_at	D20 vs MD20	-1.324	TOP2B	topoisomerase (DNA) II beta 180kDa	DNA topological change
867.1.S1_at	D20 vs MD20	-1.536	CSNK2B	casein kinase II beta subunit	Cell cycle
Cytoskeleton organization					
9540.1.S1_s_at		-1.221	MOSPD2	motile sperm domain containing 2	Structural molecule activity
3265.1.S1_s_at		-1.224	MAP7	microtubule-associated protein 7	Microtubule cytoskeleton organi.
13036.1.S1_at		-1.285	SORBS2	Sorbin and SH3 domain containing 2	Cytoskeletal adaptor activity
9747.1.S1_a_at	MD20	-1.299	IFT80	intraflagellar transport 80 homolog	Microtubule basal body
Affx.22790.3.S1_s_at	D20 vs MD20	-1.309	ABLIM1	actin binding LIM protein 1	Cytoskeleton organization
Endocytosis					
Affx.5354.1.S1_at		-1.246	ZFYVE20	zinc finger, FYVE domain containing 20	Endocytosis
7965.1.S1_at		-1.292	RAMP3	receptor (G prot.) activity modifying prot. 3	Receptor-mediated endocytosis
19332.1.S1_s_at	D20 vs MD20	-1.350	MIA3	melanoma inhibitory activity family, member 3	Exocytosis
7554.1.S1_s_at	D20 vs MD20	-1.468	RAB22A[2]	RAB22A, member RAS oncogene family	Endocytosis
ER to Golgi transport					
Affx.24394.1.S1_at		1.245	TIMM50	translocase of inner mitochondrial 50	Protein transport
1665.2.S1_a_at		1.232	SRPR	signal recognition particle receptor	Protein export
5523.2.S1_a_at		-1.204	MKKS	McKusick-Kaufman syndrome	Protein folding
11349.1.S1_s_at	D20 vs MD20	-1.218	GOLGA5	golgi autoantigen, golgin subfamily a, 5	Golgi organization
12408.3.S1_a_at		-1.240	ATXN3	ataxin 3	Protein processing in ER
10119.1.S1_at	D20 vs MD20	-1.295	SEC22C	SEC22 vesicle trafficking protein homolog C	ER to Golgi vesicle-med. transport
5900.3.S1_a_at		-1.320	DNAJA4[2]	DnaJ (Hsp40) homolog, subfamily A, member 4	Protein folding
10753.1.S1_at		-1.371	REEP3[2]	receptor accessory protein 3	Cellular vesicle trafficking
5586.1.S1_at	D20 vs MD20	-1.415	LOC416931	similar to vacuolar sorting protein VPS29	Protein transport
Immune system					
Affx.9422.1.S1_at		1.414	PTGER4	prostaglandin E receptor 4 (subtype EP4)	Immune response
11227.1.S1_at		1.403	BST1	bone marrow stromal cell antigen 1	Humoral immune response
4511.1.S1_at	D20 vs MD20	1.311	TCIRG1	T-cell, immune reg. 1, lysosomal V0, A3	Cellular defense response

Affx.23995.1.S1_at		1.245	BMP6	bone morphogenetic protein 6	Immune response
Affx.2640.1.S1_s_at		-1.205	NPAL3	NIPA-like domain containing 3	Immune response
6007.2.S1_a_at		-1.244	RNF128	ring finger protein 128	Neg. reg. cytokine biosynthesis
Affx.1020.4.S1_s_at	D20 vs MD20	-1.366	C5	complement component 5	Pos. reg. of chemokine secretion
956.1.S1_at		-1.442	NFIL3	nuclear factor, interleukin 3 regulated	Immune response
Lipid and fatty acid metabolism					
5289.1.S1_at	D20 vs MD20	-1.209	GPD1	Glycerol-3-phosphate dehydrogenase 1 (soluble)	Glycerophospholipid metabolism
Affx.2668.1.S1_at		-1.218	PTDSS2	phosphatidylserine synthase 2	Glycerophospholipid metabolism
Affx.5721.1.S1_at	D20 vs MD20	-1.253	LYCAT	lysocardiolipin acyltransferase	Phospholipid biosynthetic process
Affx.5556.1.S1_at	MD20	-1.256	HAO1	hydroxyacid oxidase (glycolate oxidase) 1	Fatty acid alpha-oxidation
Affx.21829.1.S1_s_at	D20 vs MD20	-1.286	ST8SIA5	ST8 α-N-acetyl-neuraminide α-2,8-sialyltransf. 5	Glycosphingolipid biosynthesis
13239.1.S1_at	D20 vs MD20	-1.301	ACAD9	acyl-Coenzyme A dehydrogenase family, 9	Beta-oxidation of fatty acyl-CoA
Affx.10598.1.S1_s_at		-1.335	EPT1	ethanolaminephosphotransferase 1	Phospholipid biosynthetic process
6352.1.S1_at	D20 vs MD20	-1.385	HADHB	hydroxyacyl-Coenzyme A dehydrogenase, beta	Fatty acid elongation
Miscellaneous					
6043.2.S1_a_at		1.738	PTTG1IP	Pituitary tumor-transforming 1 interacting prot.	Protein import into nucleus
2698.1.S1_at	MD20	1.645	MYL4	myosin, light chain 4, alkali; atrial, embryonic	Pos. reg. of ATPase activity
13504.1.S1_at	D20 vs MD20	1.563	RRM2	ribonucleotide reductase M2 polypeptide	Glutathione metabolism
Affx.3712.1.S1_s_at	D20 vs MD20	1.528	DAK	dihydroxyacetone kinase 2 homolog	Glycerolipid metabolism
18941.1.S1_at	D20 vs MD20	1.402	TMEM30C	transmembrane protein 30C	Transmembrane protein
74.1.S1_at		1.299	LMO2	LIM domain only 2 (rhombotin-like 1)	Blood vessel development
7008.1.S1_at		1.207	ALDH4A1	aldehyde dehydrogenase 4 family, member A1	Arginine and proline metabolism
8396.1.S1_s_at		-1.203	DERA	2-deoxyribose-5-phosphate aldolase homolog	Deoxyribonucleoside catabolism
Affx.24282.1.S1_at		-1.205	SRD5A3	steroid 5 alpha-reductase 3	Steroid hormone biosynthesis
Affx.9117.5.S1_s_at		-1.209	MAN1A2	mannosidase, alpha, class 1A, member 2	N-glycan biosynthesis
16674.1.S1_at		-1.210	LOC427419	similar to myosin IIIA; deafness	Microfilament motor activity
7599.3.S1_a_at		-1.217	LOC420810	similar to Im:7137941 protein	Oxidoreductase activity
Affx.12601.1.S1_s_at		-1.221	ANP32E	acidic nuclear phosphoprotein 32, E	Phosphatase inhibitor activity
17326.1.S1_s_at		-1.222	ORMDL1[3]	ORM1-like 1 / solute carrier family 40, 1	Ceramide metabolic process
10998.1.S1_s_at		-1.222	TMEM63A	transmembrane protein 63A	Nucleotide binding
16268.1.S1_at	D20 vs MD20	-1.225	AADACL2	arylacetamide deacetylase-like 2	Carboxylesterase activity
12043.2.S1_a_at	MD20	-1.226	PALMD	Palmdelphin	Regulation of cell shape
Affx.24266.1.S1_at		-1.228	TXNDC10[3]	thioredoxin domain containing 10	Thioredoxin-related transmembrane
8268.1.S1_s_at		-1.229	NDST2	N-deacetylase/N-sulfotransferase 2	Hep. sulfate proteoglycan biosynth.
19216.1.S1_at		-1.229	CREB3L2	cAMP responsive element binding prot. 3-like 2	Response to ER stress
Affx.3114.1.S1_at		-1.239	SAMD8	sterile alpha motif domain containing 8	Sphingomyelin biosynthetic process
Affx.25881.1.S1_at	MD20	-1.239	PECAM1[2]	platelet/endothelial cell adhesion molecule	Phagocytosis
Affx.10901.2.S1_s_at		-1.245	ATP7B[3]	ATPase, Cu++ transporting, beta polypeptide	Cellular copper ion homeostasis
Affx.4733.1.S1_at		-1.256	MTMR8	myotubularin related protein 8	Protein tyrosine phosphat. activity
12189.1.S1_s_at		-1.261	FAM44A	family with sequence similarity 44, member A	DNA binding
Affx.9202.1.S1_s_at	MD20	-1.266	LEPREL2	leprecan-like 2	Collagen prolyl hydroxylases
Affx.1000.1.S1_at	D20 vs MD20	-1.266	FN3KRP	fructosamine-3-kinase-related protein	Kinase activity
Affx.26494.1.S1_at		-1.271	FRRS1	ferric-chelate reductase 1	Ferric-chelate reductase activity
12485.1.S1_at	D20 vs MD20	-1.271	AQP9	aquaporin 9	Purine transport
4639.1.S2_at		-1.273	TRAM1	translocation associated membrane protein 1	Prot. targeting to membrane
18094.1.S1_s_at	D20 vs MD20	-1.285	QSOX2	quiescin Q6 sulfhydryl oxidase 2	Cell redox homeostasis
Affx.9633.1.S1_s_at	MD20	-1.301	B4GALT6	UDP-Gal:βGlcNAc β 1,4- galactosyltransf., 6	Galactosyltransferase activity
Affx.9832.1.S1_at		-1.322	FKTN	Fukutin	Muscle organ development
Affx.8434.1.S1_at	MD20	-1.344	DAO	D-amino-acid oxidase	Arginine and proline metabolism
15933.1.S1_s_at		-1.368	ATP11C	ATPase, class VI, type 11C	ATP biosynthetic process
Affx.2216.2.S1_s_at	MD20	-1.387	LOC427438	hypothetical LOC427438	NAD+ kinase activity
Affx.22856.1.S1_at		-1.401	ZADH1	zinc bind. alcohol dehydrogenase, dom. cont. 1	Metabolism of prostaglandins
Affx.6124.1.S1_at	D20 vs MD20	-1.405	C1GALT1	C1, N-acetylgalactosamine3β-galactosyltransf. 1	Angiogenesis
7576.1.S1_at	D20 vs MD20	-1.413	ACAA1	acetyl-Coenzyme A acyltransferase 1	Cellular lipid metabolic process
Affx.8023.1.S1_s_at	D20 vs MD20	-1.459	ALDH5A1	aldehyde dehydrogenase 5 family, member A1	Gamma-aminobutyric acid catabolism
10268.1.S1_at		-1.472	SAMM50	sorting and assembly machinery component 50	Prot. import into mitochondria
RNA processing					
10535.1.S1_s_at	D20 vs MD20	1.203	CNOT1	CCR4-NOT transcription complex, subunit 1	RNA splicing and nuclear export
2565.2.S1_s_at		-1.202	UTP6	UTP6, small subunit processome component	RNA processing
Affx.24313.1.S1_at		-1.239	FIP1L1	FIP1 like 1 (S. cerevisiae)	mRNA processing
2272.1.S1_s_at		-1.242	SBDS	Shwachman-Bodian-Diamond syndrome	rRNA processing

12650.1.A1_s_at		-1.271	FMR1	fragile X mental retardation 1	mRNA transport
5501.2.S1_a_at	D20 vs MD20	-1.277	DUS1L	dihydrouridine synthase 1-like (S. cerevisiae)	tRNA processing
5254.1.S1_at		-1.279	UTP11L	UTP11-like	rRNA processing
Signal Transduction					
5969.2.S1_a_at		1.339	PAQR7	progestin/adipoQ receptor family member VII	Progesterone receptor
17590.1.S1_s_at		1.271	CLIC2	chloride intracellular channel 2	Signal transduction
Affx.20844.1.S1_s_at		1.249	SGK196	protein kinase-like protein SgK196	Protein phosphorylation
Affx.4973.2.S1_s_at		-1.216	TRAF6	TNF receptor-associated factor 6	Signal transduction
12808.1.S1_at		-1.218	STYX	serine/threonine/tyrosine interacting protein	Protein dephosphorylation
Affx.6416.1.S1_at		-1.224	TRHDE	thyrotropin-releasing hormone degrading enzyme	Signal transduction
8903.2.S1_s_at		-1.243	PPP2R5C	protein phosphatase 2, regulatory subunit B', γ	Signal transduction
Affx.3462.1.S1_at		-1.266	NTSR1	neurotensin receptor 1 (high affinity)	G-prot. coupled receptor sig. path.
12434.1.S1_at		-1.278	MAP3K7	mitogen-activated protein kinase kinase kinase 7	Signal transduction
Affx.7330.2.S1_s_at	D20 vs MD20	-1.316	SCYL2[2]	SCY1-like 2 (S. cerevisiae)	Protein phosphorylation
Affx.592.5.S1_s_at	D20 vs MD20	-1.349	RALGPS1	Ral GEF with PH domain/SH3 binding motif 1	Intracellular signaling pathway
Affx.22298.1.S1_s_at	D20 vs MD20	-1.483	GFRA3	GDNF family receptor alpha 3	Signal transduction
Transcription and translation					
16053.1.S1_at		-1.214	RCOR3	REST corepressor 3	Transcription regulation
19758.1.S1_s_at	MD20	-1.217	BRWD1	bromodomain and WD repeat dom. cont. 1	Transcription regulation
Affx.22889.1.S1_s_at		-1.229	EIF3A	eukaryotic translation initiation factor 3, A	Translation initiation factor
Affx.2415.1.S1_s_at		-1.380	LRRFIP1[2]	leucine rich repeat (in FLII) interacting protein 1	Negative regulation of transcription
13932.1.S1_s_at		-1.447	BACH1[2]	BTB and CNC homology 1	Transcription regulation
Transport					
6731.2.S1_a_at	D20 vs MD20	1.248	SLC23A1	solute carrier family 23, 1	L-ascorbic acid transport
Affx.3745.2.A1_s_at	D20 vs MD20	-1.225	SLC22A13	solute carrier family 22, 13	Transmembrane transport
Affx.26624.2.S1_s_at		-1.232	TRPM7	transient receptor potential cation channel, M, 7	Calcium ion transport
17536.1.S1_s_at	D20 vs MD20	-1.261	KCTD9	potassium channel tetramerisation dom. cont. 9	Potassium ion transport
3105.1.S1_at		-1.310	SLC38A2[2]	solute carrier family 38, member 2	Amino acid transport
9198.1.S1_at		-1.388	RHCE	Rh blood group, D antigen	Transmembrane transport
9672.1.S1_at	D20 vs MD20	-1.410	C14orf68	chromosome 14 open reading frame 68	Transmembrane transport
Affx.10113.2.S1_s_at	MD20	-1.493	CPNE3	copine III	Vesicular-mediated transport
Ubiquitin proteolytic pathway					
13366.1.S1_s_at		-1.218	ZNRF2	Zinc and ring finger 2	E3 ubiquitin ligase
17645.1.S1_s_at	D20 vs MD20	-1.242	TRIP12	thyroid hormone receptor interactor 12	Ubiquitin mediated proteolysis
1883.2.S1_a_at	D20 vs MD20	-1.311	USP15	ubiquitin specific peptidase 15	Protein deubiquitination
Affx.9219.1.S1_at	MD20, B vs C	-1.367	USP5	ubiquitin specific peptidase 5 (isopeptidase T)	Reg. proteaso. ubiquitin-dep. catabol.
Xenobiotic metabolism					
1128.2.S1_a_at		1.841	SOD3	superoxide dismutase 3, extracellular	Removal of superoxide radicals
617.1.S1_at	D20 vs MD20	1.516	CYP1A1	cytochrome P450, family 1, subfamily A, 1	Respnse to drug
618.1.S1_at	D20 vs MD20	1.447	CYP1A4	cytochrome P450 1A4	Xenobiotic enzyme (chicken)
Affx.25940.1.S1_at		1.208	PAPSS2	3'-phosphoadenosine 5'-phosphosulfate synth. 2	Sulfation detoxification
10745.2.S1_s_at	D20 vs MD20	-1.298	CYP4V2	cytochrome P450, family 4, subfamily V, 2	Response to stimulus

Other groups: gene expression regulated in other comparisons (table 9.2, 9.3).

Fc: Fold change

$P < 0.05$

N: Control 7, D20 8 samples.

[2,3]: Number of probe sets for same gene

B vs C: D20 vs MD20

Table 9.2: 2nd Deoxynivalenol experiment; significantly regulated genes in MD20

Probe Set ID, Gga.	Other groups	Fc	Gene symbol	Gene title	Gene function
Actin cytoskeleton					
2698.1.S1_at	D20	1.776	MYL4	myosin, light chain 4, alkali; atrial, embryonic	Actin filament binding
8364.1.S1_at		-1.201	PTPN12	protein tyrosine phosphatase, non-receptor 12	Actin cytoskeleton

4058.1.S2_s_at		-1.239	LOC396473[2]	Myristoylated alanine-rich C kinase substrate	Actin cytoskeleton
18929.1.S1_s_at	D20 vs MD20	-1.239	WIPF1	WAS/WASL interacting protein family, 1	Actin cytoskeleton organization
Affx.13048.1.S1_at		-1.263	PLS3[2]	plastin 3 (T isoform)	Actin-binding protein
2409.1.S1_at		-1.274	WDR1	WD repeat domain 1	Disassembly of actin filaments
Affx.23003.1.S1_at		-1.341	PALLD	palladin, cytoskeletal associated protein	Reg. actin cytoskeleton organization
12385.1.S1_at	D20 vs MD20	-1.631	LIMA1[2]	LIM domain and actin binding 1	Actin filament bundle assembly
Apoptosis					
Affx.8419.3.S1_s_at	D20 vs MD20	1.238	SOS1	son of sevenless homolog 1 (Drosophila)	Apoptosis
Affx.9034.1.S1_at	D20 vs MD20	1.210	LRDD	leucine-rich repeats and death dom. cont.	Apoptosis
Affx.4468.2.S1_s_at	D20 vs MD20	-1.249	ACOX2	acyl-Coenzyme A oxidase 2, branched chain	Bile acid metabolic process
Affx.8434.1.S1_at	D20	-1.253	DAO	D-amino-acid oxidase	Arginine and proline metabolism
Affx.10484.1.S1_at	D20 vs MD20	-1.266	ATAD2	ATPase family, AAA domain containing 2	Assembly transcrip. coreg. compl.
Affx.12075.1.S1_at		-1.269	LZIC	leucine zipper and CTNNBIP1 domain containing	Beta-catenin binding
11829.1.S1_s_at		-1.296	FGGY	FGGY carbohydrate kinase domain containing	Carbohydrate metabolic process
Affx.11407.1.S1_s_at		-1.302	YWHAE	tyrosine 3-tryptophan 5-monooxygenase act. p., ε	Induction apoptosis extracellular sig.
Affx.21196.1.S1_s_at	D20 vs MD20	-1.311	RERE	arginine-glutamic acid dipeptide (RE) repeats	Apoptosis
16833.1.S1_at		-1.312	TP53INP1	tumor protein p53 inducible nuclear protein 1	Apoptosis
10204.1.S1_s_at	D20 vs MD20	-1.317	CASP1	caspase 1, apoptosis-related cysteine peptidase	Apoptosis
Affx.24739.1.S1_at		-1.319	JPH1[2]	junctophilin 1	Calcium ion transport into cytosol
6127.1.S1_at	D20 vs MD20	-1.331	CABC1	chaperone, ABC1 activity of bc1 complex hom.	Cell death
Affx.21316.1.S1_s_at	D20 vs MD20	-1.348	AMD1[2]	adenosylmethionine decarboxylase 1	Arginine and proline metabolism
Affx.23333.1.S1_at		-1.594	SPATA4	spermatogenesis associated 4	Apoptosis
19230.1.S1_s_at		-1.661	EAF2	ELL associated factor 2	Apoptosis
Affx.9237.1.S1_s_at	D20 vs MD20	-4.521	PPP1R3C[2]	protein phosphatase 1, regulatory 3C	Carbohydrate metabolic process
Cholesterol and lipid metabolism					
Affx.9142.1.S1_at	D20 vs MD20	2.398	PPARGC1A	PPARγ, coactivator 1α	Energy metabolism
Affx.2762.2.S1_at	D20 vs MD20	1.609	CRAT[2]	carnitine acetyltransferase	Fatty acid β-oxidation
1568.1.S1_at	D20 vs MD20	1.302	PPAP2B	Phosphatidic acid phosphatase type 2B	Lipid metabolic process
18859.1.S1_at		-1.211	ACLY	ATP citrate lyase	Cholesterol biosynthesis
Affx.8801.1.S1_at	D20 vs MD20	-1.240	MVK	mevalonate kinase (mevalonic aciduria)	Cholesterol biosynthesis
13412.1.S1_at		-1.241	OSBPL2	Oxysterol binding protein-like 2	Cellular cholesterol homeostasis
7273.1.S1_at		-1.241	TLCD1	TLC domain containing 1	Lipid metabolism
13515.1.S1_at		-1.255	LBR[2]	lamin B receptor	Cholesterol biosynthesis
7130.1.S1_at		-1.298	OSBPL10	oxysterol binding protein-like 10	Hepatic lipogenesis
16889.1.S1_at	D20 vs MD20	-1.324	MYLIP	myosin regulatory light chain interacting protein	LDLR ubiquitination
Affx.12964.1.S1_s_at	D20 vs MD20	-1.336	LSS[2]	lanosterol synthase	Cholesterol biosynthesis
2847.2.S1_a_at		-1.399	MMAB	methylmalonic aciduria cblB type	Cobalamin biosynthetic process
4949.1.S1_at	D20 vs MD20	-1.419	SREBF2	sterol regulatory element binding TF 2	Cholesterol metabolic process
12018.1.S1_at		-1.420	HSD3B7	hydroxy-Δ5-steroid dehydrog. 3β-, Δ-isomerase 7	Primary bile acid biosynthesis
Affx.12935.1.S1_s_at		-1.424	DHCR24	24-dehydrocholesterol reductase	Cholesterol biosynthesis
Affx.11821.1.S1_s_at	D20 vs MD20	-1.457	SC4MOL	sterol-C4-methyl oxidase-like	Cholesterol biosynthesis
8851.1.S1_a_at		-1.466	IDI1[2]	isopentenyl-diphosphate delta isomerase 1	Cholesterol biosynthetic process
Affx.5556.1.S1_at	D20	-1.569	HAO1	hydroxyacid oxidase (glycolate oxidase) 1	Fatty acid alpha-oxidation
Affx.6950.2.S1_s_at	D20 vs MD20	-1.616	ANGPTL3	Angiopoietin-like protein 3	Cholesterol metabolic process
Affx.12414.1.S1_s_at		-1.684	HMGCR[2]	3-hydroxy-3-methylglutaryl-Coenzyme A reduct.	Cholesterol biosynthetic process
9630.1.S1_s_at	D20 vs MD20	-1.695	LDLR	low density lipoprotein receptor	Cholesterol homeostasis
15541.1.S1_at		-1.741	DHCR7	7-dehydrocholesterol reductase	Cholesterol biosynthetic process
7215.1.S1_at	D20 vs MD20	-1.806	HSD17B7[2]	hydroxysteroid (17-beta) dehydrogenase 7	Steroid biosynthetic process
9949.1.S1_at		-1.832	NSDHL	NAD(P) dependent steroid dehydrogenase-like	Steroid biosynthesis
17055.1.S1_at		-1.914	SCD	stearoyl-CoA desaturase (delta-9-desaturase)	Fatty acid biosynthetic process
Affx.13153.1.S1_at	D20 vs MD20	-2.384	AACS[2]	acetoacetyl-CoA synthetase	Fatty acid metabolic process
8817.1.S1_s_at		-2.936	FDFT1	farnesyl-diphosphate farnesyltransferase 1	Steroid biosynthetic process
Cell cycle, growth and differentiation					
14699.1.S1_at	D20 vs MD20	1.377	SH3D19	SH3 domain containing 19	Cellular membrane organization
12849.1.S1_at	D20 vs MD20	1.329	SMOC2	SPARC related modular calcium binding 2	Angiogenesis
123.1.S1_at	D20 vs MD20	1.279	MUSK	muscle, skeletal, receptor tyrosine kinase	Multicell. organismal development
4032.1.S1_at		1.231	NRP1	neuropilin 1	Cell differentiation
15874.1.S1_at		1.208	KLHDC3	kelch domain containing 3	Cell cycle
15747.2.S1_s_at		-1.202	PDS5B	PDS5, homolog B	Cell proliferation
5613.1.S1_at	D20 vs MD20	-1.202	BANP	BTG3 associated nuclear protein	Cell cycle
Affx.12255.1.S1_at	D20 vs MD20	-1.204	PHC1	polyhomeotic homolog 1 (Drosophila)	Cellular proliferation

Affx.11588.1.S1_s_at		-1.207	SKP2	S-phase kinase-associated protein 2 (p45)	Cell proliferation
Affx.5444.1.S1_s_at		-1.210	SEPT6	septin 6	Cell cycle
13174.1.S1_at		-1.211	TUBD1	Tubulin, delta 1	Cell differentiation
9835.1.S1_a_at		-1.214	DYNLT3	dynein, light chain, Tctex-type 3	Cell cycle
12092.1.S1_at		-1.221	FAM20C	family with sequence similarity 20, C	Cell differentiation
Affx.21168.1.S1_s_at		-1.222	LIN9	lin-9 homolog (C. elegans)	Cell cycle
3200.1.S1_s_at		-1.229	SMC4	structural maintenance of chromosomes 4	Mitotic chromosome condensation
8373.1.S1_at		-1.229	RAPGEF3	Rap guanine nucleotide exchange factor (GEF) 3	Cell proliferation
Affx.25881.1.S1_s_at	D20	-1.233	PECAM1	platelet/endothelial cell adhesion molecule	Cell recognition
12043.1.S1_at	D20	-1.242	PALMD[2]	palmdelphin	Regulation of cell shape
1139.1.S1_at	D20 vs MD20	-1.243	MESDC1	Mesoderm development candidate 1	ES cell growth / differentiation
2117.1.S1_at		-1.249	WDR5	WD repeat domain 5	Chromatin modification
10765.1.S1_s_at		-1.251	LYAR	Ly1 antibody reactive homolog (mouse)	Cell growth regulation
7449.1.S1_at		-1.255	LOC416335	apical protein 2	Cell morphogenesis
7812.2.S1_a_at	D20 vs MD20	-1.256	ORC6L	origin recognition complex, subunit 6 like	DNA replication
Affx.9514.1.S1_at	D20 vs MD20	-1.262	FRK	fyn-related kinase	Cell differentiation
2363.2.S1_a_at	D20 vs MD20	-1.285	NCAPD2	non-SMC condensin I complex, subunit D2	Mitotic chromosome condensation
Affx.23401.4.S1_s_at		-1.297	DOCK7	dedicator of cytokinesis 7	Cell differentiation
Affx.24202.1.S1_at	D20 vs MD20	-1.302	RANBP10[2]	RAN binding protein 10	Microtubule cytoskeleton organizat.
Affx.5725.1.S1_at		-1.302	ACTL6A[2]	actin-like 6A	Chromatin remodeling
15959.1.S1_at		-1.305	ARL3	ADP-ribosylation factor-like 3	Cell cycle
Affx.24391.1.S1_at	D20 vs MD20	-1.312	PDS5A	PDS5, homolog A	Cell cycle
8155.1.S1_at	D20 vs MD20	-1.315	AR	androgen receptor	Cell growth
Affx.11089.1.S1_at		-1.333	KLHDC5	kelch domain containing 5	Microtubule related
7164.1.S1_at	D20 vs MD20	-1.340	LOC421110	similar to DNA replication initiator protein	DNA replication
4606.1.S1_at		-1.357	SMARCA2	SWI/SNF related, MRC, A2	Chromatin remodeling
5674.2.S1_s_at	D20 vs MD20	-1.364	APPBP1	amyloid beta precursor protein binding protein 1	Cell cycle progression
Affx.11696.1.S1_s_at	D20 vs MD20	-1.372	MCM2	minichromosome maintenance complex com. 2	DNA replication
4370.2.S1_s_at	D20 vs MD20	-1.423	MCM6	minichromosome maintenance complex com. 6	DNA replication
Affx.4473.1.S1_at	D20	-1.473	FAM107A	family with sequence similarity 107, A	Regulation of cell growth
11844.1.S1_s_at	D20 vs MD20	-1.553	KLF10[2]	Kruppel-like factor 10	Cell proliferation
Affx.25068.3.S1_s_at	D20 vs MD20	-1.583	TBC1D8[2]	TBC1 domain family, member 8	Pos. reg. of cell proliferation
17114.1.S1_s_at	D20 vs MD20	-1.748	HLF[3]	hepatic leukemia factor	Multicellular organismal development
3982.1.S1_at		-1.844	INHBA	inhibin, beta A	Cell cycle arrest
Affx.5197.1.S1_s_at	D20 vs MD20	-3.477	BUB1[2]	budding uninhibited by benzimidazoles 1 hom.	Mitotic cell cycle checkpoint
Cell junction					
6597.1.S1_at		-1.299	SDC1	syndecan 1	Cell adhesion molecule
Affx.20887.1.S1_at		-1.303	ZYX	zyxin	Cell adhesion
1722.1.S1_at		-1.337	TMEM49	transmembrane protein 49	Cell-cell adhesion, invasion
4661.2.S1_at	D20 vs MD20	-1.338	TUBA1C	tubulin, alpha 1c	Gap junction
Affx.21308.1.S1_at		-1.394	DCBLD1[2]	Discoidin, CUB and LCCL domain containing 1	Cell adhesion
Affx.21790.1.S1_s_at	D20 vs MD20	-1.823	GJA1	gap junction protein, alpha 1, 43kDa	Gap junction assembly
862.1.S1_at		-1.921	GJC1	gap junction protein, gamma 1, 45kDa	Gap junction
DNA repair					
5209.1.S2_at		-1.230	PARP1	poly (ADP-ribose) polymerase family, member 1	DNA repair and cell death
1017.1.S1_s_at		-1.236	DDA1[2]	DET1 and DDB1 associated 1	DNA repair
1820.2.S1_s_at		-1.250	RPA2[2]	replication protein A2, 32kDa	DNA repair
Affx.12131.1.S1_s_at	D20 vs MD20	-1.258	POLE2	polymerase (DNA directed), epsilon 2	DNA repair
14972.1.S1_at	D20 vs MD20	-1.430	LRRC3B	leucine rich repeat containing 3B	DNA repair
Endocytosis and vesicular transport					
Affx.6689.1.S1_at	D20 vs MD20	1.392	STON2	stonin 2	Endocytosis
Affx.11670.1.S1_s_at		1.295	EHD3	EH-domain containing 3	Endosome-to-Golgi transport
18955.1.S1_at		-1.218	VPS24	vacuolar protein sorting 24 homolog	Endocytosis
Affx.12840.1.S1_s_at		-1.233	RAB5A	RAB5A, member RAS oncogene family	Endocytosis
Affx.26182.1.S1_s_at		-1.319	LOC415406	similar to neural precursor cell expressed 4	Endocytosis
Affx.24985.2.S1_s_at	D20 vs MD20	-1.426	ITSN2[2]	intersectin 2	Endocytosis
Immune system					
Affx.2639.1.S1_at		1.289	SIGIRR	single Ig and toll-interleukin 1 receptor domain	Innate immune response
7298.1.S1_at	D20 vs MD20	-1.217	C1QA	complement component 1, q A	Innate immune response
Affx.11512.1.S1_at		-1.217	TRAFD1	TRAF-type zinc finger domain containing 1	Neg. reg. of innate immune response

3723.1.S1_at		-1.232	PLA2G4A[2]	phospholipase A2, group IVA	Regulation of inflammatory reponse
Affx.23449.1.S1_at		-1.243	IFIH1	interferon induced with helicase C domain 1	Innate immune response
4039.1.S1_s_at	D20 vs MD20	-1.255	IGF2BP3	insulin-like growth factor 2 mRNA BP 3	Reg. cytokine biosynthetic process
11597.1.S1_s_at		-1.256	STAT1	signal transducer and activator of transcription 1	Cytokine-mediated signaling pathway
12009.1.S1_at	D20 vs MD20	-1.256	TICAM1	toll-like receptor adaptor molecule 1	dsRNA sensing
Affx.22137.1.S1_at	D20 vs MD20	-1.270	PXK	PX domain containing serine/threonine kinase	Inflammatory response
1148.1.S1_at	D20 vs MD20	-1.285	ST6GAL1[2]	ST6 β-galactosamide alpha-2,6-sialyltranferase 1	Humoral immune response
4884.1.S1_at		-1.356	ANXA1	annexin A1	Inflammatory response
19.1.S1_at		-1.373	IL1B	interleukin 1, beta	Cytokine
7841.1.S1_at	D20 vs MD20	-1.384	AGPAT2	1-acylglycerol-3-phosphate O-acyltransferase 2	Positive regulation of cytokine prod.
4285.1.S1_at	D20 vs MD20	-1.451	CEBPB	CCAAT/enhancer binding protein (C/EBP),β	Immune response
Affx.21915.1.S1_at		-1.454	IFIT5	IFN-induced protein with tetratricopep. repeats 5	Inflammatory response
131.1.S1_at		-1.575	MX1	myxovirus resistance 1	Defense response
Affx.21842.2.S1_s_at		-1.905	GAL6	Gal 6	Antimicrobial peptide
2190.2.S1_a_at	D20 vs MD20	-1.949	SOCS2	suppressor of cytokine signaling 2	JAK-STAT cascade
Insulin signaling					
4447.1.S1_at	D20 vs MD20	7.761	PCK1	phosphoenolpyruvate carboxykinase 1 (soluble)	Response to insulin stimulus
Affx.8324.1.S1_s_at		-1.299	GRB10	growth factor receptor-bound protein 10	Insulin receptor signaling pathway
9876.2.S1_a_at	D20 vs MD20	-1.324	PHPT1	phosphohistidine phosphatase 1	Dephosphorylation
Affx.10567.1.S1_s_at		-1.364	PHKA2[2]	phosphorylase kinase, alpha 2 (liver)	Insulin signaling pathway
11822.1.S1_at	D20 vs MD20	-2.046	PPP1R3B	protein phosphatase 1, regulatory 3B	Insulin signaling pathway
Miscellaneous					
Affx.3960.1.S1_at	D20 vs MD20	1.828	TMEM86A	transmembrane protein 86A	Transmembrane protein
4124.1.S1_at		1.786	LOC395933	sulfotransferase	Sulfotransferase activity
11193.1.S1_s_at		1.633	DMGDH	Similar to dimethylglycine dehydrogenase	Choline metabolic process
3352.1.S1_at	D20 vs MD20	1.408	WIPI1	WD repeat domain, phosphoinositide interact. 1	Multiprotein assembly
3226.1.S1_at		1.400	CRAMP1L	Crm, cramped-like (Drosophila)	DNA binding protein
674.1.S1_at	D20 vs MD20	1.339	GCH1	GTP cyclohydrolase 1	Dopamine biosynthetic process
5635.1.S1_at		1.326	ABCC5	ATP-binding cassette, sub-family C, member 5	Multi-drug resistance
17156.1.S1_at		1.300	ALG14	asparagine-linked glycosylation 14 homolog	Protein folding
Affx.882.2.S1_s_at	D20 vs MD20	1.296	COX10[3]	COX10 homolog, cytochrome c oxidase	Cellular respiration
10224.1.S1_at	D20 vs MD20	1.292	TRIM65[2]	tripartite motif-containing 65	Metal ion binding
2318.1.A1_at	D20 vs MD20	1.273	KIAA0247	KIAA0247	Monocyte chemoattractant prot.-1
Affx.9445.1.S1_s_at	D20 vs MD20	1.260	NNT[2]	nicotinamide nucleotide transhydrogenase	NADPH producing for radical detox.
15866.2.S1_a_at		1.258	SLC25A12	solute carrier family 25, member 12	L-glutamate / aspartate transport
Affx.12190.2.A1_s_at		1.246	COG1	component of oligomeric golgi complex 1	Golgi organisation
Affx.10287.1.S1_at		1.240	BACE2	beta-site APP-cleaving enzyme 2	Alzheimers's disease
8353.1.S1_at		-1.202	CSTB	cystatin B (stefin B)	Neg. reg. of peptidase activity
8926.2.S1_s_at		-1.203	AZIN1	antizyme inhibitor 1	Polyamine biosynthetic process
2840.1.S1_at		-1.203	B4GALT1	UDP-Gal:betaGlcNAc β1,4- galactosyltransf., 1	Oligosaccharide biosynthetic process
Affx.12174.1.S1_s_at		-1.205	PEPD	peptidase D	M.etallocarboxypeptidase activity
Affx.11753.1.S1_at		-1.211	PDK1	pyruvate dehydrogenase kinase, isozyme 1	Reg. acetyl-CoA synth. F. pyruvate
7671.2.S1_s_at	D20 vs MD20	-1.212	GMPS	guanine monphosphate synthetase	Drug metabolism
12290.1.S1_at		-1.216	XRCC2	X-ray repair com. def. repair, hamster cells 2	Homologous recombination
Affx.25706.1.S1_s_at		-1.217	ERCC4	excision repair cross-complementing, compl. 4	Nucleotide excision repair
Affx.24053.1.S1_s_at		-1.229	DENND5B	DENN/MADD domain containing 5B	Membrane portein
Affx.20499.1.S1_at		-1.230	LOC426187	similar to LOC594885 protein	Melanocyte related protein
1201.1.S1_s_at		-1.231	GYG1	glycogenin 1	Glycogen biosynthesis
Affx.24424.1.S1_at		-1.234	CDC37L1	cell division cycle 37 homolog-like 1	Cochaperone
Affx.23815.1.S1_at		-1.236	AVL9	AVL9 homolog (S. cerevisiase)	Golgi transport and sorting
Affx.11551.1.S1_s_at	D20 vs MD20	-1.236	H2AFZ	H2A histone family, member V	Nucleosome assembly
16843.1.S1_s_at	D20 vs MD20	-1.244	SEPHS1	selenophosphate synthetase 1	Selenoamino acid metabolism
Affx.11705.1.S1_at		-1.247	HAGHL	hydroxyacylglutathione hydrolase-like	Metal ion binding
Affx.13180.1.S1_s_at	D20 vs MD20	-1.250	MTAP	methylthioadenosine phosphorylase	Nucleoside metabolic process
Affx.22530.1.S1_s_at		-1.252	RNPEPL1	arginyl aminopeptidase-like 1	Aminopeptidase
9534.1.S1_at		-1.256	ZDHHC6	zinc finger, DHHC-type containing 6	Metal ion binding
Affx.12101.1.S1_s_at		-1.258	N6AMT2	N-6 adenine-specific DNA methyltransferase 2	Methyltransferase
7283.1.S1_at		-1.261	ZFAND3	Zinc finger, AN1-type domain 3	Metal ion binding
Affx.26492.1.S1_s_at		-1.263	AGL[2]	amylo-1, 6-glucosidase, 4-α-glucanotransferase	Glycogen degradation
16749.1.S1_at		-1.269	SLC25A24	Solute carrier family 25, member 24	Mitochondrial phosphate carrier
7653.2.S1_a_at		-1.279	PRR6	proline rich 6	Centromere complex assembly
4676.1.S1_s_at	D20 vs MD20	-1.284	NUBP2	nucleotide binding protein 2	Iron-sulfur protein assembly

Affx.21489.1.S1_s_at	D20 vs MD20	-1.287	TBL1XR1	transducin (beta)-like 1 X-linked receptor 1	Glycerophospholipid metabolism
Affx.9202.1.S1_s_at	D20	-1.292	LEPREL2	leprecan-like 2	Collagen prolyl hydroxylases
10299.1.S1_a_at		-1.302	ACN9	ACN9 homolog (S. cerevisiae)	Regulation of gluconeogenesis
8083.1.S1_at		-1.303	ORMDL1	ORM1-like 1 (S. cerevisiae)	Ceramide metabolic process
Affx.12152.1.S1_s_at		-1.306	FAM53A	family with sequence similarity 53, A	Nuclear protein
11920.2.S1_a_at		-1.311	RG9MTD2	RNA guanine-9-methyltransferase dom. Cont. 2	Methyltransferase
16542.1.S1_at	D20 vs MD20	-1.315	TMEM106B[3]	transmembrane protein 106B	Transmembrane protein
2412.1.S1_at		-1.315	PPID	peptidylprolyl isomerase D (cyclophilin D)	Protein folding
3867.1.S1_at	D20 vs MD20	-1.318	PRNP	prion protein (p27-30)	Cellular copper ion homeostasis
16294.1.S1_a_at		-1.324	SEC61A2	Sec61 alpha 2 subunit (S. cerevisiae)	Transmembrane transport
Affx.23122.1.S1_s_at	D20 vs MD20	-1.332	JMJD2A	jumonji domain containing 2A	Histone demethylation
5742.1.S1_s_at	D20 vs MD20	-1.341	HIP1R	huntingtin interacting protein 1 related	Receptor-mediated endocytosis
737.1.S1_at	D20 vs MD20	-1.342	SLC2A2	solute carrier family 2, member 2	Glucose transport
939.1.S1_at		-1.343	LOC395310	prominin-like protein	Cell surface protein
Affx.26453.1.S1_at	D20 vs MD20	-1.352	DAK[2]	dihydroxyacetone kinase 2 homolog	Glycerolipid metabolism
9747.1.S1_a_at	D20	-1.353	IFT80	intraflagellar transport 80 homolog	Cytoskeleton
Affx.20694.1.S1_s_at	D20 vs MD20	-1.361	ORMDL1[2]	ORM1-like 1 (S. cerevisiae)	Ceramide metabolic process
Affx.8317.3.S1_at		-1.362	DDC[2]	dopa decarboxylase	Cellular AA / derivative metabolism
4564.2.S1_a_at	D20 vs MD20	-1.369	FMO6[2]	flavin containing monooxygenase 6	Monooxygenase activity
Affx.348.2.S1_s_at	D20 vs MD20	-1.402	ETNK2	ethanolamine kinase 2	Glycerophospholipid metabolism
11760.1.S1_at		-1.413	TMEM103	transmembrane protein 103	Transmembrane protein
7662.1.S1_at	D20 vs MD20	-1.414	NUDT15	nudix-type motif 15	Metal ion binding
2353.1.S1_at	D20 vs MD20	-1.414	TIPARP	TCDD-inducible poly(ADP-ribose) polymerase	Androgen metabolic process
5625.1.S1_s_at		-1.425	PGP	phosphoglycolate phosphatase	Phosphoglycolate phosphatase act.
8095.1.S1_at	D20 vs MD20	-1.445	MPP6	membrane protein, palmitoylated 6	Protein complex assembly
12131.1.S1_at	D20 vs MD20	-1.448	MID1IP1	MID1 interacting protein 1	Acetyl-CoA carboxylase α inhibitor
Affx.23773.1.S1_s_at		-1.449	MLH1	mutL homolog 1, colon cancer, nonpolyposis 2	Mismatch repair
1645.1.S1_s_at		-1.466	UBXN2B	UBX domain protein 2B	Golgi and ER biogenesis
Affx.24796.1.S1_at		-1.499	WWP1	WW domain cont. E3 ubiquitin prot. ligase 1	Protein ubiquitination
Affx.9633.1.S1_s_at	D20	-1.503	B4GALT6[2]	UDP-Gal:betaGlcNAc β1,4- galactosyltransf., 6	Galactosyltransferase activity
14188.1.S1_at	D20 vs MD20	-1.516	ASMTL[2]	acetylserotonin O-methyltransferase-like	Melatonin biosynthetic process
Affx.20416.1.S1_at	D20 vs MD20	-1.519	LOC427550	similar to Pyruvate dehydrog. Phosphat. cat. 2	Reg. acetyl-CoA synt. from pyruvate
Affx.4296.1.S1_at		-1.545	PISD	phosphatidylserine decarboxylase	Phospholipid biosynthetic process

RNA processing

Affx.240.5.S1_s_at		-1.202	HNRNPM	heterogeneous nuclear ribonucleoprotein M	RNA splicing
Affx.25712.2.S1_s_at	D20 vs MD20	-1.203	NUP93	nucleoporin 93kDa	mRNA transport
Affx.3493.1.S1_at		-1.206	MSI2	musashi homolog 2 (Drosophila)	RNA binding protein
9658.1.S1_at		-1.206	NOLA2	nucleolar protein family A, member 2	rRNA processing
6100.2.S1_a_at		-1.211	LSM6	LSM6 homolog, U6 small nuclear RNA assoc.	RNA splicing
8127.1.S1_at		-1.211	PUS7	pseudouridylate synthase 7 homolog	RNA binding
16753.1.S1_s_at	D20 vs MD20	-1.215	TARDBP	TAR DNA binding protein	RNA splicing
1182.1.S1_s_at	D20 vs MD20	-1.217	MOV10	Moloney leukemia virus 10, homolog (mouse)	mRNA cleavage for gene silencing
12366.1.S1_at		-1.224	NUP37	nucleoporin 37kDa	mRNA transport
Affx.9018.1.S1_at		-1.224	NUP50	nucleoporin 50kDa	mRNA transport
3409.2.S1_a_at		-1.230	SNRPE	small nuclear ribonucleoprotein polypeptide E	RNA splicing
Affx.22512.1.S1_s_at		-1.244	PDCD11	programmed cell death 11	rRNA maturation and 18S generation
Affx.26068.1.S1_s_at		-1.246	HNRNPH3	heterogeneous nuclear ribonucleoprotein H3	RNA processing
7927.1.S1_at	D20 vs MD20	-1.261	MED9	mediator of RNA polymerase II transcription 9	RNA polymerase II activation
9585.1.S1_s_at		-1.272	POP4	processing of precursor 4, ribonuclease P/MRP	Processing precursor RNA
13135.1.S1_at		-1.298	ZCRB1	Zinc finger CCHC-type and RNA binding motif 1	RNA splicing
Affx.2291.1.S1_s_at		-1.414	RNASEL	ribonuclease L	28S rRNA cleavage
3356.1.S1_at		-1.438	NHP2L1	NHP2 non-histone chromosome protein 2-like 1	RNA splicing

Signaling

Affx.24290.1.S1_at	D20 vs MD20	2.149	DUSP16[2]	dual specificity phosphatase 16	Dephosporylation of JNK1/2 MAPK14
Affx.24302.1.S1_at	D20 vs MD20	1.637	MC5R[2]	melanocortin 5 receptor	PI3K-regulated activation of ERK1/2
12952.1.S1_at	D20 vs MD20	1.392	INPP5B	inositol polyphosphate-5-phosphatase, 75kDa	Phosphatidylinositol signaling
Affx.20364.1.S1_at		1.289	AGPAT6	1-acylglycerol-3-phosphate O-acyltransferase 6	Signal transduction/lipid synthesis
Affx.8419.5.S1_s_at		1.246	LOC431192	similar to alternate SOS1	Signal transduction
Affx.3808.1.S1_at		-1.207	MAPK8	mitogen-activated protein kinase 8	JNK cascade
5646.1.S1_at		-1.208	DGKZ	diacylglycerol kinase, zeta 104kDa	Activation of protein kinase C activity
7296.1.S1_at		-1.210	IFNGR2	interferon gamma receptor 2	Jak-STAT signaling pathway
1550.1.S1_s_at		-1.212	TFG	TRK-fused gene	Pos. reg. I-κB kinase/NF-κB cascade

2771.1.S1_at		-1.215	SUFU	suppressor of fused homolog (Drosophila)	Hedgehog signaling pathway
19634.1.S1_s_at	D20 vs MD20	-1.218	NKIRAS1	NFKB inhibitor interacting Ras-like 1	I-kappaB kinase/NF-kappaB cascade
10418.1.S1_at	D20 vs MD20	-1.225	GNG10	Guanine nucleotide binding protein, gamma 10	Signal transduction
3763.1.S1_at		-1.225	SNF1LK2	SNF1-like kinase 2	Intracellular protein kinase cascade
2993.2.S1_a_at		-1.237	NCOR1	nuclear receptor co-repressor 1	Negative regulation of JNK cascade
16454.1.S1_at		-1.250	KLHL2	kelch-like 2, Mayven (Drosophila)	Intracellular protein transport
11842.1.S1_s_at	D20 vs MD20	-1.256	SYDE2	Synapse defective 1, Rho GTPase, homolog 2	Activation of Rho GTPase activity
1584.1.S1_s_at		-1.284	LOC419104	similar to USP6NL protein	Regulation of Rab GTPase activity
Affx.26288.1.S1_s_at		-1.302	MAPK6	mitogen-activated protein kinase 6	MAPK pathway
Affx.6110.1.S1_s_at	D20 vs MD20	-1.311	ARHGAP11A	Rho GTPase activating protein 11A	Signal transduction
Affx.12195.1.S1_s_at	D20 vs MD20	-1.386	YWHAH	tyrosine 3-/tryptophan 5-monooxygenase act. η	Glucocorticoid receptor signaling
12556.1.S1_at	D20 vs MD20	-1.393	ERRFI1	ERBB receptor feedback inhibitor 1	Negative regulation of EGFR activity
17662.1.S1_s_at		-1.393	RALGPS2	Ral GEF with PH domain and SH3 binding 2	Small GTPase med. signal transd.
Affx.343.1.S1_at	D20 vs MD20	-1.403	CNKSR1	connector enhancer of kinase suppressor Ras 1	Ras protein signal transduction
6368.1.S1_at	D20 vs MD20	-1.497	DUSP14	dual specificity phosphatase 14	Dephosphorylation ERK1/2 / JNK1/2
3387.1.S1_at		-1.582	ADRA1B	adrenergic, alpha-1B-, receptor	Cell-cell signaling
10870.1.S1_a_at		-1.591	ASB5	ankyrin repeat and SOCS box-containing 5	Intracellular signaling pathway
5622.1.S1_at	D20 vs MD20	-1.755	PIK3R1[2]	Phosphoinositide-3-kinase, regulatory 1	Growth hormone receptor signaling
Transcription					
7228.1.S1_at	D20 vs MD20	1.914	CMBL	carboxymethylenebutenolidase homolog	Xenobiotic metabolism
Affx.24402.1.S1_at	D20 vs MD20	1.761	RBPJ	recombination signal BP for Igκ J region	Notch signaling pathway
Affx.21033.1.S1_at	D20 vs MD20	1.661	BCL9[3]	B-cell CLL/lymphoma 9	Wnt receptor signaling pathway
13311.1.S1_s_at	D20 vs MD20	1.394	VPS26B	vacuolar protein sorting 26 homolog B	Vacuolar transport
Affx.5382.1.S1_s_at	D20 vs MD20	-1.208	COBRA1	cofactor of BRCA1	Negative regulation of transcription
3412.1.S1_at C	D20 vs MD20	-1.211	CITED4	Cbp/p300-interacting transactivator 4	Transcription regulation
11449.1.S1_at	D20 vs MD20	-1.247	MLLT6	Myeloid/lymphoid/mixed-lineage leukemia 6	Regulation of transcrption
5440.1.S1_s_at	D20 vs MD20	-1.325	BCL7A[2]	B-cell CLL/lymphoma 7A	Negative regulation of transcription
10787.1.S1_at	D20 vs MD20	-1.355	BATF3	basic leucine zipper TF, ATF-like 3	Transcription regulation
7664.1.S1_at		-1.402	PRRG1	proline rich Gla (G-carboxyglutamic acid) 1	Vit. K transmembrane protein
Affx.24840.1.S1_at	D20,D20vsMD20	-1.434	BRWD1[2]	bromodomain and WD repeat dom. Cont. 1	Transcription regulation
Affx.10113.2.S1_s_at	D20	-1.445	CPNE3	copine III	Vesicle-mediated transport
Affx.9219.1.S1_at	D20,D20vsMD20	-1.556	USP5[2]	ubiquitin specific peptidase 5 (isopeptidase T)	Ubiquitin-dep.protein catabolism
Affx.12584.1.S1_at	D20 vs MD20	-2.030	TADA2L[3]	transcriptional adaptor 2 ADA2 homolog-like	Transcription RNA poly. II promoter
Translation					
2560.1.S1_s_at		1.382	MRPL38	mitochondrial ribosomal protein L38	Translation
8998.1.S1_at		-1.213	URB1	ribosome biogenesis 1 homolog	Ribosome biogenesis
6364.3.S1_at		-1.233	RBM19	RNA binding motif protein 19	Ribosome biogenesis
10019.1.S1_at	D20 vs MD20	-1.271	GTPBP10	GTP-binding protein 10 (putative)	Ribosome biogenesis
Affx.2464.1.S1_at		-1.285	PRMT3	protein arginine methyltransferase 3	Ribosomal subunit level regulation
4309.1.S1_at	D20 vs MD20	-1.507	SOX8	SRY (sex determining region Y)-box 8	Reg.transcript. RNA poly. II promoter
Transport					
Affx.6963.1.S1_at		1.221	KCNK1[2]	potassium channel, subfamily K, member 1	Potassium ion transport
Affx.6403.1.S1_at		-1.224	SLC29A1	solute carrier family 29, member 1	Nucleoside transporter
12550.1.S1_at	D20 vs MD20	-1.256	KCTD14	potassium channel tetramerisation dom. cont. 14	Potassium ion transport
8136.1.S1_at	D20 vs MD20	-1.266	KIFAP3	kinesin-associated protein 3	Retrograde transport Golgi to ER
7330.2.S1_a_at		-1.492	SLC9A7[3]	solute carrier family 9, member 7	K:hydrogen antiporter activity
6433.1.S1_at	D20 vs MD20	-2.095	CLDN2	claudin 2	Transport
Affx.2399.1.S1_s_at		-2.610	TRPM1[3]	transient receptor potential cation channel M1	Transport
10181.1.S1_at	D20 vs MD20	-2.853	TTR	Transthyretin	Transport

Other groups: gene expression regulated in other comparisons (table 9.1, 9.3).

Fc: Fold change

$P < 0.05$

N: Control 7, MD20 7 samples.

[2,3]: Number of probe sets for same gene

Table 9.3: 2nd Deoxynivalenol experiment; significantly regulated genes in D20 vs. MD20

Probe Set ID, Gga.	Other groups	Fc	Gene symbol	Gene title	Gene function
Apoptosis					
Affx.5200.2.S1_s_at		1.512	CFLAR[2]	CASP8 and FADD-like apoptosis regulator	Anti-apoptosis
1727.1.S1_s_at		1.444	PPIF[2]	peptidylprolyl isomerase F	Apoptosis reg.
6233.1.S1_at	D20	1.362	SON	SON DNA binding protein	Anti-apoptosis
Affx.8881.1.S1_s_at		1.290	PPP2R4	protein phosphatase 2A act., regulatory subunit 4	Pos. reg. of apoptosis
5455.2.S1_s_at		1.283	CIAPIN1	cytokine induced apoptosis inhibitor 1	Apoptosis inhibitor
15677.1.S1_at		1.228	API5	apoptosis inhibitor 5	Anti-apoptosis
Affx.9034.1.S1_at	MD20	1.215	LRDD	leucine-rich repeats and death domain containing	Apoptosis
3144.1.S1_at	D20	-1.221	TM2D1	TM2 domain containing 1	Apoptosis
4955.1.S1_at		-1.266	FRZB	frizzled-related protein	Pos. reg. of apoptosis
Affx.12861.1.S1_s_at		-1.325	AATF	apoptosis antagonizing transcription factor	Apoptosis inducer
10204.1.S1_s_at	MD20	-1.376	CASP1	caspase 1, apoptosis-related cysteine peptidase	Apoptosis
Affx.21196.1.S1_s_at	MD20	-1.394	RERE	arginine-glutamic acid dipeptide (RE) repeats	Apoptosis
Cholesterol					
11817.1.S1_s_at		2.162	APOB[3]	apolipoprotein B (including Ag(x) antigen)	LDL transport
4006.2.S1_a_at		1.611	PPARA[2]	peroxisome proliferator-activated receptor alpha	Neg. reg. of cholesterol storage
Affx.2615.1.S1_at		1.536	LIPC	lipase, hepatic	LDL particle remodeling
145.1.S1_at		1.433	NR2C1	nuclear receptor subfamily 2, group C, member 1	Steroid hormone receptor activity
Affx.11717.1.S1_s_at		1.391	PGRMC2[2]	progesterone receptor membrane component 2	Steroid hormone receptor activity
16889.1.S1_at	MD20	-1.205	MYLIP	myosin regulatory light chain interacting protein	LDLR ubiquitination
14426.1.S1_at		-1.231	LIPA	lipase A, lysosomal acid, cholesterol esterase	Steroid biosynthesis
7256.1.S1_at		-1.248	SEC14L2	SEC14-like 2 (S. cerevisiae)	Reg. cholesterol biosynthesis
Affx.8801.1.S1_at	MD20	-1.260	MVK	mevalonate kinase (mevalonic aciduria)	Cholesterol biosynthesis
Affx.4468.2.S1_s_at	MD20	-1.312	ACOX2	acyl-Coenzyme A oxidase 2, branched chain	Bile acid metabolic process
Affx.6950.2.S1_s_at	MD20	-1.333	ANGPTL3	angiopoietin-like 3	Cholesterol metabolic process
7215.2.S1_a_at	MD20	-1.391	HSD17B7	hydroxysteroid (17-beta) dehydrogenase 7	Steroid biosynthetic process
Affx.11821.1.S1_s_at	MD20	-1.395	SC4MOL	sterol-C4-methyl oxidase-like	Cholesterol biosynthesis
4949.1.S1_at	MD20	-1.427	SREBF2	sterol regulatory element binding TF 2	Cholesterol metabolic process
7444.1.S1_at		-1.431	MVD	Mevalonate (diphospho) decarboxylase	Cholesterol biosynthesis
Affx.12964.1.S1_s_at	MD20	-1.441	LSS	lanosterol synthase	Cholesterol biosynthesis
9630.1.S1_s_at	MD20	-1.699	LDLR	low density lipoprotein receptor	Cholesterol homeostasis
9580.2.S1_a_at		-2.223	AKR1B1	aldo-keto reductase family 1, member B1	Steroid metabolic process
Cell junction					
Affx.21089.1.S1_s_at		1.437	LMO7	LIM domain 7	Cell-cell adhesion
Affx.1002.1.S1_s_at		1.261	STAB1	stabilin 1	Cell adhesion
Affx.26244.4.S1_s_at		1.251	PPFIA1[2]	tyrosine phosphatase, receptor f, interact. prot. A1	Cell matrix adhesion
Affx.24862.1.S1_at		1.222	IGSF5	immunoglobulin superfamily, member 5	Tight junction
Affx.12459.1.S1_at		-1.236	LGALS8	lectin, galactoside-binding, soluble, 8	Integrin-like cell interaction
2582.1.S1_at		-1.247	TGFBI	transforming growth factor, beta-induced	Neg. reg. of cell adhesion
7794.1.S1_at		-1.350	TUBB6	tubulin, beta 6	Gap junction
4661.2.S1_at	MD20	-1.522	TUBA1C[2]	tubulin, alpha 1c	Gap junction
Affx.21790.1.S1_s_at	MD20	-1.806	GJA1	gap junction protein, alpha 1, 43kDa	Gap junction assembly
6433.1.S1_at	MD20	-2.138	CLDN2	claudin 2	Tight junction
Cell cycle					
Affx.5146.1.S1_s_at		1.680	NT5C2[2]	5'-nucleotidase, cytosolic II	Const. composition purine/pyrimidine
15961.1.S1_s_at		1.614	CEP76	centrosomal protein 76kDa	Reg. of centriole replication
Affx.5036.1.S1_s_at		1.607	HORMAD2	HORMA domain containing 2	Mitosis
13995.1.S1_s_at		1.597	HDAC10	histone deacetylase 10	Chromatin modification
Affx.26544.1.S1_at		1.560	CTR9	Ctr9, Paf1/RNA polymerase II complex comp.	Histone H2B ubiquitination
Affx.26688.2.S1_s_at		1.552	PDE8A	phosphodiesterase 8A	Cyclic nucleotide metabolic process
Affx.11962.1.S1_s_at		1.543	PSME4[3]	proteasome activator subunit 4	Cell differentiation
Affx.10236.2.S1_at		1.515	TTC3[2]	tetratricopeptide repeat domain 3	Neg. reg. cell morphogenesis involved
857.2.S1_s_at		1.491	EGFR	epidermal growth factor receptor	Pos. reg. of cell proliferation
Affx.24643.1.S1_at		1.448	SESN1	sestrin 1	Neg. reg. of cell proliferation
Affx.12567.1.S1_s_at		1.395	RB1CC1	RB1-inducible coiled-coil 1	Cell cycle
13486.1.S1_s_at		1.366	FBF1	Fas (TNFRSF6) binding factor 1	Cell polarization
1331.2.S1_a_at	D20	1.312	TOP2B	topoisomerase (DNA) II beta 180kDa	DNA topological change

1908.1.S1_at		1.295	MEF2A[2]	myocyte enhancer factor 2A	Cell differentiation
669.1.S1_at		1.278	MET	met proto-oncogene	Cell proliferation
11155.1.S1_s_at		1.272	STAG2	stromal antigen 2	Cell cycle
Affx.26161.1.S1_at	D20	1.261	CGNL1	cingulin-like 1	Cell junction
Affx.13056.1.S1_at		1.248	E2F5	E2F transcription factor 5, p130-binding	Cell cycle regulator
10852.1.S1_s_at	D20	1.248	LRRC8A	leucine rich repeat containing 8 family, member A	Cell differentiation
10003.1.S1_at	D20	1.233	MKL2	MKL/myocardin-like 2	Cell differentiation
Affx.21938.1.S1_at		1.223	KLHL22	kelch-like 22 (Drosophila)	Cell division
11678.4.S1_a_at		1.215	SPG21	spastic paraplegia 21, maspardin	Cell death
804.1.S1_at		1.206	TGFBR3	transforming growth factor, beta receptor III	Cell growth
Affx.26541.1.S1_at		-1.200	GFER	growth factor, augmenter of liver regeneration	Augmenter of liver regeneration
5613.1.S1_at	MD20	-1.202	BANP	BTG3 associated nuclear protein	Cell cycle
1984.2.A1_a_at		-1.203	MAD2L1	MAD2 mitotic arrest deficient-like 1 (yeast)	Mitotic cell cycle checkpoint
Affx.12144.1.S1_s_at		-1.204	NEIL1	nei endonuclease VIII-like 1 (E. coli)	Base excision repair
4908.2.S1_s_at		-1.205	ANP32B	acidic nuclear phosphoprotein 32 family, B	Histone acetylation
8784.1.S1_at		-1.218	GATAD2A	GATA zinc finger domain containing 2A	DNA methylation
Affx.9514.1.S1_at	MD20	-1.219	FRK	fyn-related kinase	Cell differentiation
5913.1.S1_a_at		-1.219	MSL1	male-specific lethal 1 homolog (Drosophila)	Chromatin modification
2479.1.S1_at		-1.219	ING1	inhibitor of growth family, member 1	Neg. reg. of cell growth
7812.2.S1_a_at	MD20	-1.223	ORC6L	origin recognition complex, subunit 6 like (yeast)	DNA replication
Affx.12131.1.S1_s_at	MD20	-1.224	POLE2	polymerase (DNA directed), epsilon 2	DNA repair
1069.1.S1_at		-1.231	SMARCD2	SWI/SNF related, matrix assoc., AR chromatin D2	Chromatin remodeling complex
5864.1.S1_at		-1.245	BZW1	basic leucine zipper and W2 domains 1	Cell growth
17585.2.S1_a_at		-1.248	HDGF	Hepatoma-derived growth factor	Cell proliferation
639.1.S1_at		-1.249	S100A11	S100 calcium binding protein A11	Neg. reg. of cell proliferation
14492.1.S1_at		-1.252	NT5C1B	5'-nucleotidase, cytosolic IB	Nucleotide metabolic process
Affx.24391.1.S1_at	MD20	-1.252	PDS5A	PDS5, reg. of cohesion maintenance, homolog A	Cell cycle
6306.1.S1_at		-1.256	BAMBI	BMP/activin membrane-bound inhibitor homolog	Pos. reg. of cell proliferation
2010.1.S1_at	D20	-1.256	TOP2A	topoisomerase (DNA) II alpha 170kDa	DNA ligation
1551.2.S1_a_at	D20	-1.263	BIRC5	baculoviral IAP repeat-containing 5 (survivin)	Cell division
Affx.23665.1.S1_at		-1.269	H2B-VII[2]	similar to histone H2B	Nucleosome assembly
2207.1.S1_at		-1.273	RBL1	Retinoblastoma-like 1 (p107)	Cell cycle
Affx.11648.1.S1_at		-1.273	BCL6	B-cell CLL/lymphoma 6 (zinc finger protein 51)	Neg. reg. S phase of mitotic cell cycle
Affx.11551.1.S1_s_at	MD20	-1.274	H2AFZ	H2A histone family, member V	Nucleosome assembly
854.1.S1_a_at		-1.277	LGALS1	lectin, galactoside-binding, soluble, 1	Autocrine Neg. growth factor
Affx.12255.1.S1_at	MD20	-1.278	PHC1	polyhomeotic homolog 1 (Drosophila)	Cellular proliferation
6127.1.S1_at	MD20	-1.284	CABC1	chaperone, ABC1 activity of bc1 complex hom.	Cell death
5209.1.S1_a_at		-1.300	PARP1	poly (ADP-ribose) polymerase family, member 1	DNA repair and cell death
5674.2.S1_s_at	MD20	-1.302	APPBP1	amyloid beta precursor protein binding protein 1	Cell cycle progression
Affx.12961.1.S1_at	D20	-1.311	TPX2	TPX2, microtubule-associated, homolog	Cell proliferation
11321.1.S1_at		-1.313	S100A10	S100 calcium binding protein A10	Cell cycle progression
4514.2.S1_s_at		-1.321	RRM1	ribonucleotide reductase M1 polypeptide	Deoxyribonucleotide biosynthesis
2844.2.S1_s_at	D20	-1.324	STMN1	stathmin 1	Mitotic spindle organization
12556.1.S1_at	MD20	-1.325	ERRFI1	ERBB receptor feedback inhibitor 1	Upregulated with cell growth
11746.1.S1_at	D20	-1.330	PBK	PDZ binding kinase	Mitosis
Affx.11918.1.S1_s_at		-1.339	MXD4[2]	MAX dimerization protein 4	Reg. of cell growth
Affx.4245.1.S1_at	MD20	-1.358	KLHL25	kelch-like 25 (Drosophila)	Cell division
Affx.23122.1.S1_s_at	MD20	-1.365	JMJD2A	jumonji domain containing 2A	Histone demethylation
3213.1.S1_at		-1.370	E2F1	E2F transcription factor 1	Cell cycle regulator
1281.1.S1_at		-1.384	PTTG1	pituitary tumor-transforming 1	Cell cycle
Affx.1524.1.S1_s_at		-1.400	CSNK1G1	casein kinase 1, gamma 1 /// similar to KIAA0101	Growth and morphogenesis
3146.1.S1_at		-1.453	CCNB2	cyclin B2	Cell cycle regulator
14972.1.S1_at	MD20	-1.454	LRRC3B	leucine rich repeat containing 3B	DNA repair
8155.1.S1_at	MD20	-1.490	AR	androgen receptor	Cell growth
726.2.S1_a_at	D20	-1.491	CDC2	cell division cycle 2, G1 to S and G2 to M	Cell cycle
2363.2.S1_a_at	MD20	-1.512	NCAPD2[2]	non-SMC condensin I complex, subunit D2	Mitotic chromosome condensation
Affx.11578.1.S1_s_at	D20	-1.522	KPNA2	karyopherin alpha 2	Cell cycle
18957.1.S1_at	D20	-1.534	MKI67	antigen identified by monoclonal antibody Ki-67	Cell proliferation
Affx.24837.1.S1_at	MD20	-1.538	KLF10[2]	Kruppel-like factor 10	Cell proliferation
5542.1.S1_at	MD20	-1.557	TBC1D8[3]	TBC1 domain family, member 8	Pos. reg. of cell proliferation
Affx.12177.1.S1_s_at		-1.568	MCM5	minichromosome maintenance complex comp. 5	DNA replication
4370.2.S1_s_at	MD20	-1.627	MCM6	minichromosome maintenance complex comp. 6	DNA replication
4066.1.S1_at	D20	-1.664	SMC2	structural maintenance of chromosomes 2	Cell division
Affx.25730.1.S1_at	MD20	-1.672	HLF[4]	hepatic leukemia factor	Multicellular organismal development

7164.1.S1_at	MD20	-1.742	LOC421110	similar to DNA replication initiator protein	DNA replication
13504.1.S1_at	D20	-1.883	RRM2	ribonucleotide reductase M2 polypeptide	Deoxyribonucleotide biosynthesis
Affx.5197.1.S1_s_at	MD20	-3.286	BUB1[2]	BUB1 budding uninhibited by benzimidazoles 1	Mitotic cell cycle checkpoint
Cytoskeleton organization					
Affx.9663.1.S1_at		2.108	CEP78	centrosomal protein 78kDa	Microtubule organizing center
Affx.6391.2.S1_s_at		1.555	ANKRD15	ankyrin repeat domain 15	Neg. reg. actin filament polymerization
Affx.26411.1.S1_s_at		1.474	EPS8L2	EPS8-like 2	Actin cytoskeletal remodeling
4602.1.S1_at		1.444	BAIAP2L1	BAI1-associated protein 2-like 1	Actin assembly
Affx.22790.3.S1_s_at	D20	1.388	ABLIM1[2]	actin binding LIM protein 1	Cytoskeleton organization
13077.1.S1_s_at		1.386	PHLDB2	pleckstrin homology-like domain,B2	Microtubule stabilization
3831.1.S2_at		1.360	PACSIN2[2]	prot. kinase C and casein kin. subst. in neurons 2	Actin cytoskeleton organization
5398.1.S1_at		1.291	AGAP1	ArfGAP with GTPase, ankyrin repeat/PH dom. 1	Membrane traffic / actin cytoskeleton
5935.1.S1_s_at		1.222	SNTB1	syntrophin, beta 1	Cytoskeletal protein
10622.1.S1_at		1.203	GNA13	guanine nucleotide binding protein, alpha 13	Reg. of actin cytoskeleton
2409.1.S2_at		-1.214	WDR1	WD repeat domain 1	Disassembly of actin filaments
18929.1.S1_s_at	MD20	-1.221	WIPF1	WAS/WASL interacting protein family, member 1	Actin cytoskeleton organization
1297.1.S1_at		-1.223	LMNB2	lamin B2	Structural protein
4372.1.S1_s_at		-1.231	LOC770103	similar to Tropomyosin alpha-3 chain	Microtubule binding
19211.1.S1_s_at		-1.232	LASP1	LIM and SH3 protein 1	Cytoskeletal organization
Affx.24202.1.S1_at	MD20	-1.253	RANBP10[2]	RAN binding protein 10	Microtubule cytoskeleton organizat.
2105.2.S1_a_at		-1.275	MPRIP[2]	myosin phosphatase Rho interacting protein	Myosin light chain phosphorylation
4098.1.S1_at		-1.286	CNP	2',3'-cyclic nucleotide 3' phosphodiesterase	Microtubule cytoskeleton organizat.
3.1.S1_at		-1.398	LIMK1	LIM domain kinase 1	Reg. of actin cytoskeleton
4674.2.S1_at		-1.410	PHACTR4	phosphatase and actin regulator 4	Phosphatase and actin regulator
Affx.10663.1.S1_at		-1.418	ARSE	arylsulfatase E (chondrodysplasia punctata 1)	Skeletal system development
12385.1.S1_at	MD20	-1.501	LIMA1[2]	LIM domain and actin binding 1	Actin filament bundle assembly
3179.2.S1_at		-1.526	TUBA1C	similar to Chain A, Tubulin-Colchicine-Vinblastine	Microtubule-based movement
Endocytosis					
Affx.23990.1.S1_at		1.647	CCDC53[3]	coiled-coil domain containing 53	WASH complex endosome sorting
7554.1.S1_s_at	D20	1.614	RAB22A[2]	RAB22A, member RAS oncogene family	Endocytosis
Affx.10921.1.S1_s_at		1.390	SPG20	spastic paraplegia 20 (Troyer syndrome)	Endosomal trafficking
9995.1.S1_at		1.380	CROT	carnitine O-octanoyltransferase	Peroxisome
Affx.6689.1.S1_at	MD20	1.352	STON2	stonin 2	Reg. of endocytosis
Affx.4775.3.S1_s_at		1.350	LOC422316[2]	similar to receptor tyrosine kinase flk-1/VEGFR-2	Endocytosis
Affx.24844.1.S1_at		1.345	LRP12	low density lipoprotein-related protein 12	Endocytosis
Affx.22723.1.S1_at		1.287	EHBP1	EH domain binding protein 1	Endocytic trafficking
Affx.13170.1.S1_s_at		-1.206	CLTA	clathrin, light chain (Lca)	Endocytosis
5899.1.S1_s_at		-1.223	FAM125B	family with sequence similarity 125, member B	Endocytosis
Affx.5933.1.S1_at		-1.258	PEX1	peroxisome biogenesis factor 1	Peroxisome organization
5742.1.S1_s_at	MD20	-1.300	HIP1R[2]	huntingtin interacting protein 1 related	Receptor-mediated endocytosis
Affx.24985.2.S1_s_at	MD20	-1.334	ITSN2	intersectin 2	Endocytosis
Energy metabolism					
Affx.9032.5.S1_s_at		1.644	ATP8A1[2]	ATPase, aminophospholipid transport.,I, 8A1	ATP biosynthetic process
13239.1.S1_at	D20	1.549	ACAD9	acyl-Coenzyme A dehydrogenase family, 9	Beta-oxidation of fatty acyl-CoA
16268.1.S1_at	D20	1.287	AADACL2	arylacetamide deacetylase-like 2	Carboxylesterase activity
Affx.21381.2.S1_s_at	MD20	1.229	COX10	COX10 homolog	Cellular respiration
Affx.20134.1.S1_s_at		-1.238	GPI[2]	glucose phosphate isomerase	Carbohydrate metabolic process
Affx.9237.2.S1_at	MD20	-4.885	PPP1R3C[2]	protein phosphatase 1, regulatory subunit 3C	Carbohydrate metabolic process
ER and Golgi transport					
10119.1.S1_at	D20	1.733	SEC22C	SEC22 vesicle trafficking protein homolog C	ER to Golgi vesicle-med. transport
7523.1.S1_s_at		1.591	FNDC3A[2]	fibronectin type III domain containing 3A	Related to Golgi apparatus
9772.1.S1_s_at		1.572	FN1[3]	fibronectin 1	ER-Golgi intermediate compartment
12740.1.S1_s_at		1.457	MON2	MON2 homolog (S. cerevisiae)	Golgi to endosome transport
13311.1.S1_s_at	MD20	1.435	VPS26B	vacuolar protein sorting 26 homolog B	Vacuolar transport
Affx.26281.3.S1_s_at		1.429	ARFGEF2[2]	ADP-ribosylation G nucleotide-exchang. fact. 2	Intracellular vesicular trafficking
19332.1.S1_s_at	D20	1.390	MIA3	melanoma inhibitory activity family, member 3	Exocytosis
Affx.9460.1.S1_at		1.377	GOPC	golgi asso. PDZ and coiled-coil motif containing	ER-Golgi intermediate compartment
14699.1.S1_at	MD20	1.335	SH3D19	SH3 domain containing 19	Post-Golgi vesicle-med. transport
123.1.S1_at	MD20	1.332	MUSK	muscle, skeletal, receptor tyrosine kinase	Cellular vesicle trafficking
12603.1.S1_at		1.308	TRIM23	tripartite motif-containing 23	Formation of intracellular vesicles

9738.2.S1_s_at		1.295	MPPE1	metallophosphoesterase 1	ER to Golgi vesicle-med. transport
17220.1.S1_at		1.236	TOM1L2	target of myb1-like 2 (chicken)	Intracellular protein transport
11349.1.S1_s_at	D20	1.236	GOLGA5	golgi autoantigen, golgin subfamily a, 5	Golgi organization
Affx.12603.1.S1_s_at		-1.214	VPS39	vacuolar protein sorting 39 homolog	Protein transport
8136.1.S1_at	MD20	-1.242	KIFAP3	kinesin-associated protein 3	Retrograde transport Golgi to ER
Affx.21452.2.S1_s_at	MD20	-1.317	MESDC1[2]	mesoderm development candidate 1	ES cell growth / differentiation
Affx.11696.1.S1_s_at	MD20	-1.627	MCM2	minichromosome maintenance complex comp. 2	Cellular vesicle trafficking
Fatty acid metabolism					
10416.1.S1_s_at		3.020	CPT1A[2]	carnitine palmitoyltransferase 1A (liver)	Fatty acid beta-oxidation
Affx.2211.1.S1_s_at		2.881	ABCA12	ATP-binding cassette, sub-family A, member 12	Lipid transport
Affx.9142.1.S1_at	MD20	2.487	PPARGC1A	peroxisome proliferator-activated receptor γ, 1α	Energy metabolism
16444.1.S1_at		2.469	PDK4	pyruvate dehydrogenase kinase, isozyme 4	Reg. acetyl-CoA biosynthetic process
Affx.3994.1.S1_s_at		2.104	PANK1[2]	pantothenate kinase 1	Biosynthesis of coenzyme A
2537.2.S1_at		1.895	HMGCL	3-hydroxymethyl-3-methylglutaryl-Coen. A lyase	Acyl-CoA metabolic process
19214.1.S1_at	D20	1.748	ST8SIA5[2]	ST8 α-N-acetyl-neuraminide α-2,8-sialyltransf. 5	Glycosphingolipid biosynthesis
Affx.2762.2.S1_at	MD20	1.716	CRAT[2]	carnitine acetyltransferase	Fatty acid beta-oxidation
18942.1.S1_at		1.627	ACSL1[2]	acyl-CoA synthetase long-chain family member 1	Reg. of fatty acid oxidation
7576.1.S1_at	D20	1.593	ACAA1	acetyl-Coenzyme A acyltransferase 1	Cellular lipid metabolic process
6352.1.S1_at	D20	1.581	HADHB	hydroxyacyl-Coenzyme A dehydrogenase, β	Fatty acid elongation
Affx.8291.1.S1_at		1.519	LOC770453	hypothetical protein LOC770453	Lipid metabolic process
Affx.25584.4.S1_s_at		1.450	PDPR	pyruvate dehydrogen. phosphatase reg. subunit	Reg. acetyl-CoA biosynthetic process
Affx.23330.1.S1_at	MD20	1.414	PPAP2B[2]	phosphatidic acid phosphatase type 2B	Lipid metabolic process
Affx.22014.2.S1_s_at		1.395	EHHADH[3]	enoyl-CoA hydrat./3-hydroxyacyl CoA dehydrogen.	Fatty acid metabolic process
11390.1.S1_at		1.376	PECI	peroxisomal D3,D2-enoyl-CoA isomerase	Fatty acid metabolism
8980.1.S1_s_at		1.345	ACAD11	acyl-Coenzyme A dehydrogenase family, 11	Acyl-CoA dehydrogenase activity
Affx.11677.1.S1_at		1.318	ACOX1	acyl-Coenzyme A oxidase 1, palmitoyl	Fatty acid beta-oxidation pathway
Affx.5499.1.S1_s_at		1.239	PNPLA7	patatin-like phospholipase domain containing 7	Lipid metabolic process
5289.1.S1_at	D20	1.228	GPD1	Glycerol-3-phosphate dehydrogenase 1	Glycerophospholipid metabolism
4712.1.S1_a_at		-1.224	OXCT1	3-oxoacid CoA transferase 1	Cellular ketone body metabolic proc.
Affx.25875.1.S1_s_at		-1.225	ALDH9A1	aldehyde dehydrogenase 9 family, member A1	Fatty acid metabolism
8365.1.A1_at		-1.239	GUSB	glucuronidase, beta	Glycosaminoglycan degradation
Affx.20416.1.S1_at	MD20	-1.485	PDP2	pyruvate dehydrogenase phosphatase isoen. 2	Reg. acetyl-CoA biosynthetic process
Affx.3712.1.S1_s_at	MD20, D20	-1.513	DAK[3]	dihydroxyacetone kinase 2 homolog	Glycerolipid metabolism
Affx.348.2.S1_s_at	MD20	-1.553	ETNK2	ethanolamine kinase 2	Glycerophospholipid metabolism
Affx.13153.1.S1_at	MD20	-2.402	AACS	acetoacetyl-CoA synthetase	Fatty acid metabolic process
Immune response					
11127.1.S1_s_at		1.914	ABHD2[2]	abhydrolase domain containing 2	Response to wounding
Affx.4846.4.S1_s_at		1.842	FOXP1	forkhead box P1	Pos. reg. of immunoglobulin prod.
Affx.22594.1.S1_s_at		1.704	SPINT1	serine peptidase inhibitor, Kunitz type 1	Hepatic injury
Affx.1020.4.S1_s_at	D20	1.365	C5	complement component 5	Pos. reg. of chemokine secretion
10876.1.S1_s_at		1.348	HIPK2	homeodomain interacting protein kinase 2	Apoptosis induction by intracellular sig.
13641.1.S1_s_at		1.270	COLEC11	collectin sub-family member 11	Host-defense
Affx.10882.1.S1_at		1.220	CYSLTR2	cysteinyl leukotriene receptor 2	Immune response
2225.1.S1_at		1.210	IL6ST	interleukin 6 signal transducer	Cytokine receptor complex
12009.1.S1_at	MD20	-1.203	TICAM1	toll-like receptor adaptor molecule 1	dsRNA sensing
4491.1.S1_s_at		-1.209	LSP1	lymphocyte-specific protein 1	Cellular defense response
8873.1.S1_a_at		-1.216	NMI	N-myc (and STAT) interactor	Inflammatory response
7427.1.S1_at		-1.216	HBXIP	hepatitis B virus x interacting protein	Response to virus
7298.1.S1_at	MD20	-1.225	C1QA	complement component 1, q, A chain	Innate immune response
4039.1.S1_s_at	MD20	-1.251	IGF2BP3	insulin-like growth factor 2 mRNA binding prot. 3	Reg. of cytokine biosynthetic process
1148.1.S1_at	MD20	-1.276	ST6GAL1	ST6 β-galactosamide α-2,6-sialyltranferase 1	Humoral immune response
12662.1.S1_at		-1.287	STAT1[2]	signal transducer and activator of transcription 1	Cytokine-mediated signaling pathway
4285.1.S1_at	MD20	-1.294	CEBPB	CCAAT/enhancer binding protein (C/EBP), beta	Immune response
536.1.S1_a_at		-1.297	OASL	2'-5'-oligoadenylate synthetase-like	Immune response
Affx.22137.1.S1_at	MD20	-1.303	PXK	PX domain containing serine/threonine kinase	Inflammatory response
4511.1.S1_at	D20	-1.314	TCIRG1	T-cell, immune regulator 1, ATPase	Cellular defense response
7841.1.S1_at	MD20	-1.371	AGPAT2	1-acylglycerol-3-phosphate O-acyltransferase 2	Pos. reg. of cytokine production
3252.1.S1_at		-1.624	TLR5	toll-like receptor 5	Innate immune response
9806.1.S1_at	MD20	-1.918	A2LD1	AIG2-like domain 1	Immune response
Insulin signaling					
4447.1.S1_at	MD20	7.339	PCK1	phosphoenolpyruvate carboxykinase 1 (soluble)	Response to insulin stimulus

19491.1.S1_at		2.541	IGFBP1	insulin-like growth factor binding protein 1	Insulin receptor signaling pathway
1107.1.S1_at		1.831	PTPN2[2]	protein tyrosine phosphatase, non-receptor type 2	Insulin receptor signaling pathway
3580.2.S1_a_at		1.519	INSR	insulin receptor	Insulin receptor signaling pathway
Affx.12752.1.S1_at		1.237	PPP2R2D	protein phosphatase 2, regulatory subunit B, δ	Akt regulation
Affx.23069.1.S1_s_at		1.202	GAB1	GRB2-associated binding protein 1	Insulin receptor signaling pathway
Affx.1551.1.S1_at		-1.220	IGFALS	insulin-like growth factor bind. prot., acid labile sub.	Insulin-like growth factor binding
2170.1.S1_at		-1.266	GRB2	growth factor receptor-bound protein 2	Insulin receptor signaling pathway
11822.1.S1_at	MD20	-1.922	PPP1R3B	protein phosphatase 1, regulatory subunit 3B	Insulin signaling pathway
Metabolism					
Affx.1939.1.S1_at		2.176	RHOBTB1	Rho-related BTB domain containing 1	Small GTPase med. signal transd.
Affx.6124.1.S1_at	D20	1.577	C1GALT1	C1 glycop.N-acetylgalactosamine 3β-galactosylt. 1	Angiogenesis
9829.1.S1_at		1.533	CA5B	carbonic anhydrase VB, mitochondrial	Hydration of carbon dioxide
12952.1.S1_at	MD20	1.473	INPP5B	inositol polyphosphate-5-phosphatase	Inositol-polyphosphate 5-phosphat.
Affx.1780.1.S1_s_at		1.468	GK5	glycerol kinase 5 (putative)	Glycerol-3-phosphate metabolism
7041.3.S1_a_at		1.453	LOC422046	similar to LOC494798 protein	Oxidation reduction
3352.1.S1_at	MD20	1.416	WIPI1	WD repeat domain, phosphoinositide interacting 1	Multiprotein assembly
18778.1.S1_s_at		1.412	FAM108B1	family with sequence similarity 108, member B1	Hydrolase activity
Affx.2893.1.S1_at		1.409	ASPA	aspartoacylase (Canavan disease)	Histidine metabolism
674.1.S1_at	MD20	1.385	GCH1	GTP cyclohydrolase 1	Dopamine biosynthetic process
Affx.8023.1.S1_s_at	D20	1.385	ALDH5A1	aldehyde dehydrogenase 5 family, member A1	γ-aminobutyric acid catabolic process
Affx.11933.1.S1_s_at		1.384	NAMPT[2]	nicotinamide phosphoribosyltransferase	Nicotinamide metabolic process
1080.2.S1_s_at		1.364	MTHFD1[2]	methylenetetrahydrofolate dehydrogenase 1	Glyoxylate/dicarboxylate metabolism
12849.1.S1_at	MD20	1.349	SMOC2	SPARC related modular calcium binding 2	Angiogenesis
11167.1.S1_at		1.292	H6PD	Hexose-6-phosphate dehydrogenase	Pentose phosphate pathway
3355.1.S1_s_at		1.287	BCKDHB	branched chain keto acid dehydrogenase E1, β	Valine, leucine and isoleucine degrad.
Affx.5721.1.S1_at	D20	1.266	LYCAT	lysocardiolipin acyltransferase	Phospholipid biosynthetic process
Affx.9870.1.S1_at		1.262	GGH	gamma-glutamyl hydrolase	Folate biosynthesis
12804.1.S1_at		1.253	GPLD1	glycosylphosphatidylinositol spec. phospholip. D1	GPI anchor release
2334.1.S2_at		1.247	PHOSPHO1	phosphatase, orphan 1	Pospho-ethanolamine phosphat. act.
12100.1.S1_at		1.217	SULF2	sulfatase 2	Heparan sulfate proteoglycan metab.
Affx.11155.1.S1_at		1.209	EXT1	exostoses (multiple) 1	Elongation heparan sulfate biosynth.
Affx.7250.1.S1_at		1.201	AMDHD1	amidohydrolase domain containing 1	Histidine metabolism
Affx.22705.1.S1_at		-1.201	ALAD	aminolevulinate, delta-, dehydratase	Heme biosynthetic process
Affx.23840.1.S1_s_at		-1.206	ADH1C	alcohol dehydrogenase 1C, gamma polypeptide	Hydroxysteroids oxidation
7815.2.S1_a_at		-1.208	LOC418170	similar to aldose reductase	Fructose and mannose metabolism
Affx.10042.1.S1_at		-1.210	ZNF704	zinc finger protein 704	Zinc ion binding
959.1.S1_at		-1.218	SOX18	SRY (sex determining region Y)-box 18	Angiogenesis
5558.1.S1_at		-1.218	ALG8	asparagine-linked glycosylation 8 homolog	N-Glycan biosynthesis
1022.2.S1_a_at		-1.237	GSTZ1	glutathione transferase zeta 1	Glutathione metabolism
5250.1.S1_a_at		-1.239	HDHD3	haloacid dehalogenase-like hydrol. dom. cont. 3	Phosphoglycolate phosphatase act.
5415.2.S1_a_at		-1.251	ENOPH1	enolase-phosphatase 1	L-meth. salvage methylthioadenosine
11764.2.S1_a_at		-1.260	GCHFR	GTP cyclohydrolase I feedback regulator	Nitric oxide biosynthetic process
12514.1.S1_a_at		-1.296	GSTA1	Glutathione S-transferase class-alpha	Glutathione metabolic process
4676.1.S1_s_at	MD20	-1.302	NUBP2	nucleotide binding protein 2	Iron-sulfur protein assembly
2827.2.S1_a_at		-1.337	TIMP3	TIMP metallopeptidase inhibitor 3	Reg. membrane p. ectodom. proteol
14188.1.S1_at	MD20	-1.342	ASMTL	acetylserotonin O-methyltransferase-like	Melatonin biosynthetic process
8280.1.S1_at		-1.362	PGAM5[2]	phosphoglycerate mutase family member 5	Phosphoprotein phosphatase activity
8794.1.S1_at	MD20	-1.365	CUTA	cutA divalent cation tolerance homolog	Response to metal ion
Affx.20694.1.S1_s_at	MD20	-1.369	ORMDL1	ORM1-like 1 (S. cerevisiae)	Ceramide metabolic process
2353.1.S1_at	MD20	-1.406	TIPARP	TCDD-inducible poly(ADP-ribose) polymerase	Androgen metabolic process
8095.1.S1_at	MD20	-1.433	MPP6	membrane protein, palmitoylated 6	Protein complex assembly
2472.1.S1_at		-1.456	RDH12	retinol dehydrogenase 12	Retinol metabolsim
12131.1.S1_at	MD20	-1.474	MID1IP1	MID1 interacting protein 1	Acetyl-CoA carboxylase α inhibitor
3179.1.S2_at		-1.524	BASP1	brain abundant, membrane attached signal prot. 1	Neg. reg. gene-specific transcription
Miscellaneous					
2833.1.S1_at		2.786	SIK1	salt-inducible kinase 1	Protein phosphorylation
Affx.3960.1.S1_at	MD20	2.352	TMEM86A	transmembrane protein 86A	Transmembrane protein
Affx.1124.1.S1_at		1.910	TMEM110	transmembrane protein 110	Transmembrane protein
Affx.4741.2.S1_s_at		1.605	LOC416086	hypothetical LOC416086	Transmembrane transport
8737.1.S1_at		1.368	MAL2	mal, T-cell differentiation protein 2	Transcytosis
5567.1.S1_at		1.356	ZFAND6	zinc finger, AN1-type domain 6	Metal ion binding
Affx.24960.1.S1_at		1.319	PQLC3	PQ loop repeat containing 3	Transmembrane protein

3217.1.S1_at		1.317	MAN1B1	mannosidase, alpha, class 1B, member 1	Protein N-linked glycosylation
10224.1.S1_at	MD20	1.310	TRIM65[2]	tripartite motif-containing 65	Metal ion binding
11048.1.S1_at		1.304	SLMAP	sarcolemma associated protein	Protein folding
6250.1.S1_at		1.297	SELENBP1	selenium binding protein 1	Protein transport
18094.1.S1_s_at	D20	1.294	QSOX2	quiescin Q6 sulfhydryl oxidase 2	Cell redox homeostasis
Affx.23577.1.S1_at		1.251	SNRK	SNF related kinase	Protein phosphorylation
12957.1.S1_at		1.237	MFAP3	microfibrillar-associated protein 3	Integral to memebrane
Affx.25166.1.S1_s_at		1.230	PAN3	polyA specific ribonuclease subunit homolog	Protein phosphorylation
Affx.7330.2.S1_s_at	D20	1.228	SCYL2	SCY1-like 2 (S. cerevisiae)	Protein phosphorylation
11204.2.S1_a_at		1.226	ANKRD46	ankyrin repeat domain 46	Membrane protein
Affx.13180.1.S1_s_at	MD20	-1.220	MTAP	methylthioadenosine phosphorylase	Cysteine and methionine metabolism
Affx.20299.1.S1_s_at	MD20	-1.231	TMEM106B	transmembrane protein 106B	Transmembrane protein
7662.1.S1_at	MD20	-1.231	NUDT15	nudix-type motif 15	Metal ion binding
9156.1.S1_s_at		-1.264	TMEM57	transmembrane protein 57	Transmembrane protein
Affx.25096.1.S1_at		-1.273	TMTC4	transmembrane/tetratricopeptide repeat cont. 4	Membrane protein
3867.1.S1_at	MD20	-1.304	PRNP	prion protein (p27-30)	Cellular copper ion homeostasis
Affx.21316.1.S1_s_at	MD20	-1.352	AMD1	adenosylmethionine decarboxylase 1	Arginine and proline metabolism
Affx.13108.1.S1_at		-1.483	DPH1[2]	DPH1 homolog (S. cerevisiae)	Diphthamide synthesis
18941.1.S1_at	D20	-1.490	TMEM30C	transmembrane protein 30C	Transmembrane protein
Affx.8972.1.S1_s_at		-1.972	MANSC1	MANSC domain containing 1	Transmembrane protein
RNA processing					
10185.1.S1_at		1.588	SF3B1	Splicing factor 3b, subunit 1, 155kDa	RNA splicing
Affx.23946.1.S1_s_at		1.319	ZCCHC6	zinc finger, CCHC domain containing 6	RNA 3'-end processing
Affx.1877.1.S1_s_at		1.300	PRPF8	PRP8 pre-mRNA processing factor 8 homolog	Spliceosome
10535.1.S1_s_at	D20	1.281	CNOT1	CCR4-NOT transcription complex, subunit 1	RNA splicing and nuclear export
Affx.5692.3.S1_s_at		1.207	CWC22	CWC22 spliceosome-associated prot. Homolog	mRNA processing
Affx.23276.1.S1_at		-1.201	ZCCHC11	zinc finger, CCHC domain containing 11	RNA 3'-end processing
Affx.9364.1.S1_at		-1.201	SFRS12IP1	SFRS12-interacting protein 1	mRNA processing
9209.1.S1_at		-1.202	ASCC3L1	act. signal cointegrator 1 complex sub. 3-like 1	RNA splicing
16753.1.S1_s_at	MD20	-1.204	TARDBP	TAR DNA binding protein	RNA splicing
Affx.25538.1.S1_at		-1.210	RRP9	small subunit (SSU) processome comp., homo.	rRNA processing
Affx.25712.2.S1_s_at	MD20	-1.265	NUP93	nucleoporin 93kDa	mRNA transport
1182.1.S1_s_at	MD20	-1.271	MOV10	Moloney leukemia virus 10, homolog (mouse)	mRNA cleavage / gene silencing
Signal transduction					
6220.1.S1_a_at		2.573	ICER	ICER protein	cAMP-mediated signaling
Affx.24290.1.S1_at	MD20	1.989	DUSP16[2]	dual specificity phosphatase 16	Dephosporylation of JNK1/2 MAPK14
11339.1.S1_at		1.952	VIPR2	vasoactive intestinal peptide receptor 2	Cell-cell signaling
Affx.22298.1.S1_s_at	D20	1.910	GFRA3	GDNF family receptor alpha 3	Signal transduction
Affx.24302.1.S1_at	MD20	1.876	MC5R[2]	melanocortin 5 receptor	PI3K-regulat. downstream act. ERK1/2
Affx.24402.1.S1_at	MD20	1.762	RBPJ	recombination sig. bind. prot. Ig κ J region	Notch signaling pathway
10425.1.S1_s_at		1.604	VSNL1	visinin-like 1	Intracellular signaling pathway
10515.3.S1_a_at		1.497	CREBL2	cAMP responsive element binding protein-like 2	Signal transduction
7880.1.S1_at	MD20	1.480	BCL9	B-cell CLL/lymphoma 9	Wnt receptor signaling pathway
11825.1.S1_s_at		1.391	GK	glycerol kinase	PPAR signaling pathway
12376.1.S1_at		1.377	TBC1D5	TBC1 domain family, member 5	Reg. of Rab GTPase activity
Affx.25172.1.S1_at		1.370	SPATA13	spermatogenesis associated 13	Reg. Rho protein signal transduction
4120.1.S1_at		1.365	DUSP1	dual specificity phosphatase 1	MAPK dephosphorylation
14100.1.S1_s_at		1.345	HHIP	hedgehog interacting protein	Neg. reg. of signal transduction
Affx.24126.1.S1_at		1.344	PLXNB2[2]	plexin B2	Reg. of small GTPase med. signal
Affx.592.5.S1_s_at	D20	1.343	RALGPS1[2]	Ral GEF with PH domain and SH3 binding motif 1	Intracellular signaling pathway
9757.1.S1_at		1.328	RGS5[3]	regulator of G-protein signalling 5	Reg. G-protein coupled receptor sig.
Affx.5930.2.S1_s_at		1.300	GRK5	G protein-coupled receptor kinase 5	G-protein signaling
Affx.4808.1.S1_s_at		1.291	ARHGAP21	Rho GTPase activating protein 21	Signal transduction
755.1.S1_at		1.263	BMPR1A	bone morphogenetic protein receptor, type IA	BMP signaling pathway
11154.1.S1_s_at		1.258	STAT3	signal transducer and activator of transcription 3	Cellular response hormone stimulus
10974.1.S1_s_at		1.245	ST5	suppression of tumorigenicity 5	Pos. reg. of ERK1 and ERK2 cascade
12322.1.S1_at		1.245	NF1	neurofibromin 1	MAPK signaling pathway
Affx.5668.1.S1_at		1.243	GNB4	guanine nucleotide binding protein, β, 4	Hormone-mediated signaling pathway
Affx.25055.1.S1_s_at		1.242	INPP4A	inositol polyphosphate-4-phosphatase, type I	Phosphatidylinositol signaling system
Affx.1000.1.S1_at	D20	1.242	FN3KRP	fructosamine 3 kinase related protein	Kinase activity
Affx.6597.1.S1_at		1.241	TM4SF1	transmembrane 4 L six family member 1	Signal transduction
Affx.8419.3.S1_s_at	MD20	1.231	SOS1	son of sevenless homolog 1 (Drosophila)	Jak-STAT signaling pathway

Affx.681.2.S1_s_at		1.228	MTMR4	myotubularin related protein 4	Dephosphorylation
8896.2.S1_a_at		1.221	TSPAN9	Tetraspanin 9	Signal transduction cell development
Affx.5712.2.S1_s_at		1.203	TBC1D8B	TBC1 domain family, member 8B	Reg. of Rab GTPase activity
11493.1.S1_s_at		-1.209	GPRC5C	G protein-coupled receptor, family C, group 5, C	Effects of retinoic acid on G prot. sig.
Affx.23685.1.S1_s_at		-1.212	LIMD1	LIM domains containing 1	Signal transduction
10418.1.S1_at	MD20	-1.227	GNG10	Guanine nucleotide binding protein, gamma 10	Signal transduction
8130.1.S1_at		-1.239	SGK2	serum/glucocorticoid regulated kinase 2	Intracellular protein kinase cascade
19634.1.S1_s_at	MD20	-1.244	NKIRAS1	NFKB inhibitor interacting Ras-like 1	I-kappaB kinase/NF-kappaB cascade
18928.1.S1_s_at		-1.249	TNRC15	trinucleotide repeat containing 15	Reg. tyrosine kinase receptor signaling
Affx.21489.1.S1_s_at	MD20	-1.292	TBL1XR1	transducin (beta)-like 1 X-linked receptor 1	Canonical Wnt receptor signaling
11842.1.S1_s_at	MD20	-1.294	SYDE2	Synapse defective 1, Rho GTPase, homolog 2	Activation of Rho GTPase activity
9876.2.S1_a_at	MD20	-1.295	PHPT1	phosphohistidine phosphatase 1	Dephosphorylation
Affx.6110.1.S1_s_at	MD20	-1.309	ARHGAP11A	Rho GTPase activating protein 11A	Signal transduction
7132.1.S1_at		-1.312	CKS1B	CDC28 protein kinase regulatory subunit 1B	Reg. cyclin-dependent prot. kinase act.
Affx.12195.1.S1_s_at	MD20	-1.344	YWHAH	tyrosine 3-/tryptophan 5-monooxygenase act. ε	Glucocorticoid receptor signaling
11252.1.S1_at		-1.354	CCL19	chemokine (C-C motif) ligand 19	Cell communication
7271.1.S1_s_at		-1.383	MPP1	membrane protein, palmitoylated 1, 55kDa	Signal transduction
11814.1.S1_at		-1.422	RSU1	Ras suppressor protein 1	Ras signal transduction pathway
Affx.343.1.S1_at	MD20	-1.451	CNKSR1	connector enhancer of kinase suppressor Ras 1	Ras protein signal transduction
6368.1.S1_at	MD20	-1.491	DUSP14	dual specificity phosphatase 14	Dephosphorylation ERK1/2 / JNK1/2
2190.2.S1_a_at	MD20	-1.597	SOCS2	suppressor of cytokine signaling 2	JAK-STAT cascade
Affx.24529.1.S1_s_at	MD20	-2.123	PIK3R1[4]	phosphoinositide-3-kinase, regulatory subunit 1	GH receptor signaling pathway
Transcription					
Affx.9243.1.S1_s_at		1.595	CNOT6	CCR4-NOT transcription complex, subunit 6	CCR4-NOT core transc. reg. complex
Affx.8072.2.S1_s_at		1.562	HIVEP1[2]	HIV type I enhancer binding protein 1	Transcription reg.
19209.1.S1_at		1.401	ZMIZ1	zinc finger, MIZ-type containing 1	Reg. of transcription factors (STAT)
5574.1.S1_at		1.378	HIC2	hypermethylated in cancer 2	Neg. reg. of transcription
19390.1.S1_s_at		1.349	MED13	mediator complex subunit 13	Transcription RNA POLY II promoter
Affx.10484.1.S1_at	MD20	-1.201	ATAD2	ATPase family, AAA domain containing 2	Assembly transccoregulator compl.
6990.1.S1_s_at		-1.223	SNAPC5[2]	small nuclear RNA activating complex, polypept. 5	Transcription RNA polym. II promoter
6056.1.S1_at		-1.223	MTDH	Metadherin	Transcription RNA POLY II promoter
7448.1.S1_at		-1.245	MED24	mediator complex subunit 24	Transcription RNA POLY II promoter
11449.1.S1_at	MD20	-1.245	MLLT6	Myeloidlymphoid/mix.lineage leukemia transloc., 6	Reg. of transcrption
3412.1.S1_at	MD20	-1.258	CITED4	Cbp/p300-interacting transactivator, 4	Transcription reg.
7927.1.S1_at	MD20	-1.261	MED9	med. of RNA polymerase II transcription, 9 hom.	Transcription RNA POLY II promoter
5440.1.S1_s_at	MD20	-1.284	BCL7A	B-cell CLL/lymphoma 7A	Neg. reg. of transcription
2975.1.S1_at		-1.321	KCTD1	potassium channel tetramerisation domain cont. 1	Neg. reg. of transcription
Affx.5382.1.S1_s_at	MD20	-1.352	COBRA1	cofactor of BRCA1	Neg. reg. of transcription
4309.1.S1_at	MD20	-1.367	SOX8	SRY (sex determining region Y)-box 8	Transcription RNA POLY II promoter
Affx.24840.1.S1_at	MD20, B vs C	-1.421	BRWD1	bromodomain and WD repeat domain cont. 1	Transcription reg.
10787.1.S1_at	MD20	-1.547	BATF3	basic leucine zipper TF, ATF-like 3	Transcription reg.
Affx.12584.1.S1_at	MD20	-2.181	TADA2L[2]	transcriptional adaptor 2-like	Transcription RNA POLY II promoter
19277.1.S1_at		-2.350	HNF4A[3]	hepatocyte nuclear factor 4, alpha	Pos. reg. of gene-specific transcript.
Translation					
17491.1.S1_s_at		1.320	CPEB3[3]	cytoplasmic polyadenylation element bind. prot. 3	Ribozyme
5501.2.S1_a_at	D20	1.271	DUS1L	dihydrouridine synthase 1-like (S. cerevisiae)	tRNA processing
7179.2.A1_a_at		-1.213	MRPS18A	mitochondrial ribosomal protein S18A	Ribosomal protein
4894.1.S1_x_at		-1.231	NACA	Nascent polypeptide-assoc. complex α subunit	Translation
10019.1.S1_at	MD20	-1.263	GTPBP10	GTP-binding protein 10 (putative)	Ribosome biogenesis
Transport					
Affx.22425.1.S1_s_at		2.384	SLC16A5	solute carrier family 16, member 5	Monocarboxylic acid transport
9046.1.S1_at		2.079	SLC25A25[2]	solute carrier family 25, member 25	Phosphate carrier
Affx.3745.2.A1_s_at	D20	1.699	SLC22A13	solute carrier family 22, member 13	Urate and nicotinate transporter
Affx.2698.4.S1_s_at		1.656	ABCA8	ATP-binding cassette, sub-family A, member 8	Transport
6149.1.S1_s_at		1.424	SLCO4A1[2]	solute carrier organic anion transport. family, 4A1	Na-independent organic anion transp.
Affx.26624.6.S1_s_at		1.396	TRPM7	transient receptor potential cation channel,M7	Calcium ion transport
10277.1.S1_at		1.388	SLC5A3	Solute carrier family 5, member 3	Myo-inositol transport
Affx.13241.2.S1_s_at		1.363	SLC16A1[2]	solute carrier family 16, member 1	Mevalonate transport
Affx.13181.1.S1_s_at		1.314	ACBD5	acyl-Coenzyme A binding domain containing 5	Transport protein
17649.1.S1_s_at		1.311	SLC30A6	solute carrier family 30, member 6	Cellular efflux of zinc
17536.1.S1_s_at	D20	1.234	KCTD9	potassium channel tetramerisation dom. cont. 9	Potassium ion transport

Affx.12896.1.S1_at		1.229	SLC7A5	solute carrier family 7, member 5	Amino acid transport
12485.1.S1_at	D20	1.204	AQP9	aquaporin 9	Purine transport
1208.2.S1_a_at		1.203	SLC25A16	solute carrier family 25, member 16	Mitochondrial carrier
6731.2.S1_a_at	D20	-1.206	SLC23A1	solute carrier family 23, member 1	L-ascorbic acid transport
Affx.9075.1.S1_at		-1.231	SLC25A22	solute carrier family 25, member 22	L-glutamate transport
12550.1.S1_at	MD20	-1.247	KCTD14	potassium channel tetramerisation dom. cont. 14	Potassium ion transport
11184.1.S1_at		-1.254	KCTD12	potassium channel tetramerisation dom. cont. 12	Potassium ion transport
8627.1.S1_at		-1.296	SLC28A2	solute carrier family 28, member 2	Purine nucleoside transport
Affx.4778.1.S1_s_at		-1.303	SLC9A3R1[2]	solute carrier family 9, member 3 regulator 1	Reg. of Na:H antiporter activity
737.1.S1_at	MD20	-1.389	SLC2A2	solute carrier family 2, member 2	Glucose transport
10181.1.S1_at	MD20	-2.823	TTR	Transthyretin	Transport
Ubiquitin mediated proteolysis					
12185.1.S1_at		1.683	MMP23B	matrix metallopeptidase 23B	Proteolysis
7025.1.S1_a_at		1.527	TNFSF5IP1	tumor necrosis fact. superfamily, 5-induced prot. 1	Proteasome assembly
Affx.9300.1.S1_at		1.457	CAST	Calpastatin	Proteolysis amyloid precursor
17645.1.S1_s_at	D20	1.378	TRIP12	thyroid hormone receptor interactor 12	Ubiquitin mediated proteolysis
Affx.3350.2.S1_s_at		1.339	USP32[2]	ubiquitin specific peptidase 32	Protein deubiquitination
1883.2.S1_a_at	D20	1.330	USP15	ubiquitin specific peptidase 15	Protein deubiquitination
Affx.5735.1.S1_at		1.286	NAALADL2	N-acetylated alpha-linked acidic dipeptidase-like 2	Proteolysis
Affx.6936.1.S1_s_at		-1.200	USP1	ubiquitin specific peptidase 1	Protein deubiquitination
2832.1.S1_at		-1.202	UBE2N	ubiquitin-conjugating enzyme E2N	Protein ubiquitination
5788.2.S1_a_at	MD20	-1.289	USP5	ubiquitin specific peptidase 5 (isopeptidase T)	Proteasomal ubiquitin-dep. catabol.
Xenobiotic metabolism					
Affx.6623.3.S1_s_at		2.175	CYP4B1	cytochrome P450, family 4, subfamily B1	Drug metabolsim
Affx.8750.1.S1_at		1.814	TMEM181[2]	transmembrane protein 181	Toxin receptor
7228.1.S1_at	MD20	1.746	CMBL	carboxymethylenebutenolidase homolog	Xenobiotic metabolism
10745.2.S1_s_at	D20	1.506	CYP4V2[2]	cytochrome P450, family 4, subfamily V2	Monooxygenase activity
Affx.9445.1.S1_s_at	MD20	1.268	NNT[2]	nicotinamide nucleotide transhydrogenase	NADPH prod. for free radical detox.
7671.2.S1_s_at	MD20	-1.236	GMPS	guanine monphosphate synthetase	Drug metabolism
4564.1.S1_a_at	MD20	-1.286	FMO6P	flavin containing monooxygenase 6 pseudogene	Monooxygenase activity
617.1.S1_at	D20	-1.354	CYP1A1	cytochrome P450, family 1, subfamily A, 1	Respnse to drug
618.1.S1_at	D20	-1.735	CYP1A4	cytochrome P450 1A4	Xenobiotic enzyme (chicken)

Downregulated gene (D20 vs MD20): Expression lower in MD20 compared to D20

Other groups: gene expression regulated in other comparisons (table 9.1, 9.2).

Fc: Fold change

$P < 0.05$

N: D20 8, MD20 7 samples.

[2,3]: Number of probe sets for same gene

10. Appendix D

Figure 10.1: Principle component analysis all experiments

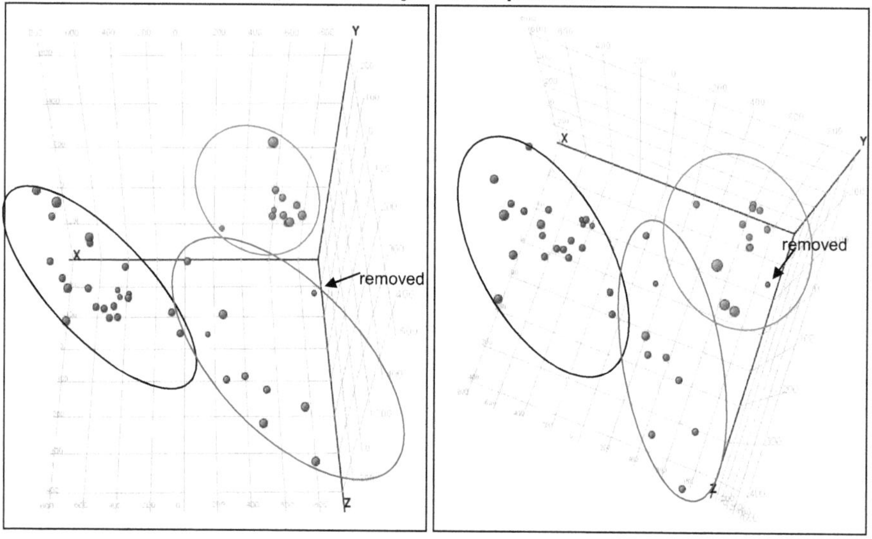

Principle component analysis from different perspectives.
Samples claret-red, red circle: Se-yeast experiment
Samples red, circle green: First DON experiment up to 5mg/kg DON
Samples blue, black circle: Second DON experiment with up to 20mg/kg DON
Sample with arrow was removed from gene expression analysis

11. List of figures

12. List of tables

MIX
Papier aus verantwortungsvollen Quellen
Paper from responsible sources
FSC® C105338

Printed by Books on Demand GmbH, Norderstedt / Germany